PHYSICAL PROBLEMS SOLVED BY
THE PHASE-INTEGRAL METHOD

This book provides a thorough introduction to one of the most efficient approximation methods for the analysis and solution of problems in theoretical physics and applied mathematics. It is written with practical needs in mind and contains a discussion of 50 problems with solutions, of varying degrees of difficulty. The problems are taken from quantum mechanics, but the method has important applications in any field of science involving second-order ordinary differential equations. The power of the asymptotic solution of second-order differential equations is demonstrated, and in each case the authors clearly indicate which concepts and results of the general theory are needed to solve a particular problem. This book will be ideal as a manual for users of the phase-integral method, as well as a valuable reference text for experienced research workers. It can also be used as a textbook for graduate students.

Nanny Fröman obtained her PhD in theoretical physics from Uppsala University. She was head of the Institute of Theoretical Physics in Uppsala for ten years and became Professor of Theoretical Physics in 1981. In 1990 she was elected a member of the Austrian Academy of Sciences, and in 1995 she was the the first woman to be elected president of the Royal Society of Sciences of Uppsala, the oldest academy in Sweden. Nanny Fröman has published numerous papers and review articles, and coauthored two previous books with Per Olof Fröman.

Per Olof Fröman obtained his PhD in theoretical physics from Uppsala University. He spent two years as a visiting researcher at the Niels Bohr Institute in Copenhagen. In 1960 he became Professor of Classical Mechanics at the Royal Institute of Technology in Stockholm and in 1964 became Professor of Theoretical Physics at Uppsala University. He was elected a member of the Austrian Academy of Sciences in 1972. Per Olof Fröman has authored and coauthored numerous scientific papers, and published two previous books with Nanny Fröman.

The authors began a joint investigation of the so-called WKB method in 1960, which resulted in their classic book *JWKB Approximation, Contributions to the Theory* from 1965. They went on to develop an efficient phase-integral method, based on the use of a phase-integral approximation of arbitrary order generated from an unspecified base function.

PHYSICAL PROBLEMS SOLVED BY THE PHASE-INTEGRAL METHOD

NANNY FRÖMAN AND PER OLOF FRÖMAN

University of Uppsala, Sweden

CAMBRIDGE
UNIVERSITY PRESS

CAMBRIDGE UNIVERSITY PRESS
Cambridge, New York, Melbourne, Madrid, Cape Town, Singapore, São Paulo

Cambridge University Press
The Edinburgh Building, Cambridge CB2 2RU, UK

Published in the United States of America by Cambridge University Press, New York

www.cambridge.org
Information on this title: www.cambridge.org/9780521812092

First published 2002
This digitally printed first paperback version 2005

A catalogue record for this publication is available from the British Library

ISBN-13 978-0-521-81209-2 hardback
ISBN-10 0-521-81209-7 hardback

ISBN-13 978-0-521-67576-5 paperback
ISBN-10 0-521-67576-6 paperback

Contents

Preface   *page* xi
1   Historical survey   1
    1.1   Development from 1817 to 1926   1
          1.1.1   Carlini's pioneering work   1
          1.1.2   The work by Liouville and Green   3
          1.1.3   Jacobi's contribution towards making Carlini's
                  work known   4
          1.1.4   Scheibner's alternative to Carlini's treatment
                  of planetary motion   4
          1.1.5   Publications 1895–1912   5
          1.1.6   First traces of a connection formula   5
          1.1.7   Publications 1915–1921   6
          1.1.8   Both connection formulas are derived in explicit form   7
          1.1.9   The method is rediscovered in quantum mechanics   7
    1.2   Development after 1926   8
2   Description of the phase-integral method   12
    2.1   Form of the wave function and the q-equation   12
    2.2   Phase-integral approximation generated from an unspecified
          base function   13
    2.3 F-matrix method   21
          2.3.1   Exact solution expressed in terms of the F-matrix   22
          2.3.2   General relations satisfied by the F-matrix   25
          2.3.3 F-matrix corresponding to the encircling of a simple
                  zero of $Q^2(z)$ 26
          2.3.4   Basic estimates   26
          2.3.5   Stokes and anti-Stokes lines   28
          2.3.6   Symbols facilitating the tracing of a wave function
                  in the complex z-plane   29

2.3.7   Removal of a boundary condition from the real z-axis
to an anti-Stokes line   30

2.3.8   Dependence of the F-matrix on the lower limit of
integration in the phase integral   32

2.3.9 F-matrix expressed in terms of two linearly independent
solutions of the differential equation   33

2.4 F-matrix connecting points on opposite sides of a well-isolated
turning point, and expressions for the wave function
in these regions   35

2.4.1   Symmetry relations and estimates of the F-matrix
elements   36

2.4.2   Parameterization of the matrix $\mathbf{F}(x_1, x_2)$ 38

2.4.2.1   Changes of α, β and γ when x_1 moves in the
classically forbidden region   40

2.4.2.2   Changes of α, β and γ when x_2 moves in the
classically allowed region   41

2.4.2.3   Limiting values of α, β and γ 42

2.4.3   Wave function on opposite sides of a well-isolated
turning point   43

2.4.4   Power and limitation of the parameterization method   45

2.5   Phase-integral connection formulas for a real, smooth,
single-hump potential barrier   46

2.5.1   Exact expressions for the wave function on both sides
of the barrier   48

2.5.2   Phase-integral connection formulas for a real barrier   50

2.5.2.1   Wave function given as an outgoing wave
to the left of the barrier   53

2.5.2.2   Wave function given as a standing wave
to the left of the barrier   54

3   Problems with solutions   59

3.1   Base function for the radial Schrödinger equation when
the physical potential has at the most a Coulomb
singularity at the origin   59

3.2   Base function and wave function close to the origin when
the physical potential is repulsive and strongly singular
at the origin   61

3.3   Reflectionless potential   62

3.4   Stokes and anti-Stokes lines   63

3.5   Properties of the phase-integral approximation along
an anti-Stokes line   66

3.6 Properties of the phase-integral approximation along a path on
     which the absolute value of exp[$iw(z)$] is monotonic
     in the strict sense, in particular along a Stokes line   66

3.7 Determination of the Stokes constants associated with the three
     anti-Stokes lines that emerge from a well isolated, simple
     transition zero   69

3.8 Connection formula for tracing a phase-integral wave function
     from a Stokes line emerging from a simple transition zero t to
     the anti-Stokes line emerging from t in the opposite direction 72

3.9 Connection formula for tracing a phase-integral wave function
     from an anti-Stokes line emerging from a simple transition
     zero t to the Stokes line emerging from t in the opposite
     direction   73

3.10 Connection formula for tracing a phase-integral wave function
     from a classically forbidden to a classically allowed region   74

3.11 One-directional nature of the connection formula for tracing
     a phase-integral wave function from a classically forbidden
     to a classically allowed region   77

3.12 Connection formulas for tracing a phase-integral wave function
     from a classically allowed to a classically forbidden region   79

3.13 One-directional nature of the connection formulas for tracing
     a phase-integral wave function from a classically allowed
     to a classically forbidden region   81

3.14 Value at the turning point of the wave function associated
     with the connection formula for tracing a phase-integral wave
     function from the classically forbidden to the classically
     allowed region   83

3.15 Value at the turning point of the wave function associated with a
     connection formula for tracing the phase-integral wave function
     from the classically allowed to the classically forbidden region 87

3.16 Illustration of the accuracy of the approximate formulas for
     the value of the wave function at a turning point   88

3.17 Expressions for the a-coefficients associated with
     the Airy functions   91

3.18 Expressions for the parameters α, β and γ when
     $Q^2(z) = R(z) = -z$ 96

3.19 Solutions of the Airy differential equation that at a fixed point
     on one side of the turning point are represented by a single,
     pure phase-integral function, and their representation on
     the other side of the turning point   98

3.20   Connection formulas and their one-directional nature
       demonstrated for the Airy differential equation   102
3.21   Dependence of the phase of the wave function in a classically
       allowed region on the value of the logarithmic derivative
       of the wave function at a fixed point x_1 in an adjacent
       classically forbidden region   105
3.22   Phase of the wave function in the classically allowed regions
       adjacent to a real, *symmetric* potential barrier, when
       the logarithmic derivative of the wave function is given
       at the centre of the barrier   107
3.23   Eigenvalue problem for a quantal particle in a broad, *symmetric*
       potential well between two *symmetric* potential barriers of
       equal shape, with boundary conditions imposed in the
       middle of each barrier   115
3.24   Dependence of the phase of the wave function in a classically
       allowed region on the position of the point x_1 in an adjacent
       classically forbidden region where the boundary condition
       $\psi(x_1) = 0$ is imposed   117
3.25   Phase-shift formula   121
3.26   Distance between near-lying energy levels in different types
       of physical systems, expressed either in terms of the
       frequency of classical oscillations in a potential well
       or in terms of the derivative of the energy with respect to
       a quantum number   123
3.27   Arbitrary-order quantization condition for a particle in
       a single-well potential, derived on the assumption
       that the classically allowed region is broad enough
       to allow the use of a connection formula   125
3.28   Arbitrary-order quantization condition for a particle in
       a single-well potential, derived without the assumption
       that the classically allowed region is broad   127
3.29   Displacement of the energy levels due to compression
       of an atom (simple treatment)   130
3.30   Displacement of the energy levels due to compression
       of an atom (alternative treatment)   133
3.31   Quantization condition for a particle in a smooth potential well,
       limited on one side by an impenetrable wall and on the other
       side by a smooth, infinitely thick potential barrier, and in
       particular for a particle in a uniform gravitational field
       limited from below by an impenetrable plane surface   137

3.32   Energy spectrum of a non-relativistic particle in a potential
        proportional to $\cot^2(x/a_0)$, where $0 < x/a_0 < \pi$ and a_0
        is a quantity with the dimension of length,
        e.g. the Bohr radius   140
3.33   Determination of a *one-dimensional*, smooth, single-well
        potential from the energy spectrum of the bound states   142
3.34   Determination of a *radial*, smooth, single-well potential
        from the energy spectrum of the bound states   144
3.35   Determination of the *radial*, single-well potential, when the
        energy eigenvalues are $-mZ^2e^4/[2\hbar^2(l+s+1)^2]$,
        where l is the angular momentum quantum number, and s is
        the radial quantum number   147
3.36   Exact formula for the normalization integral for the wave
        function pertaining to a bound state of a particle in a
        radial potential   150
3.37   Phase-integral formula for the normalized radial wave
        function pertaining to a bound state of a particle in a
        radial single-well potential   152
3.38   Radial wave function $\psi(z)$ for an s-electron in a classically
        allowed region containing the origin, when the potential
        near the origin is dominated by a strong, attractive
        Coulomb singularity, and the normalization factor is chosen
        such that, when the radial variable z is dimensionless,
        $\psi(z)/z$ tends to unity as z tends to zero   155
3.39   Quantization condition, and value of the normalized wave
        function at the origin expressed in terms of the level
        density, for an s-electron in a single-well potential with a
        strong attractive Coulomb singularity at the origin   160
3.40   Expectation value of an unspecified function $f(z)$ for
        a non-relativistic particle in a bound state   163
3.41   Some cases in which the phase-integral expectation value
        formula yields the expectation value exactly in the
        first-order approximation   166
3.42   Expectation value of the kinetic energy of a non-relativistic
        particle in a bound state. Verification of the virial theorem   167
3.43   Phase-integral calculation of quantal matrix elements   169
3.44   Connection formula for a complex potential barrier   171
3.45   Connection formula for a real, single-hump potential barrier   181
3.46   Energy levels of a particle in a smooth double-well potential,
        when no symmetry requirement is imposed   186

3.47   Energy levels of a particle in a smooth, *symmetric*,
          double-well potential   190

3.48   Determination of the quasi-stationary energy levels
          of a particle in a radial potential with a thick
          single-hump barrier   192

3.49   Transmission coefficient for a particle penetrating a real
          single-hump potential barrier   197

3.50   Transmission coefficient for a particle penetrating a real,
          symmetric, superdense double-hump potential barrier   200

References   205
Author index   209
Subject index   211

Preface

Only a few quantal problems are exactly soluble by analytical methods. In practice one has almost always to resort to an appropriate approximation method or to numerical integration of the Schrödinger equation. Among the great variety of approximate methods used in quantum mechanics, one may discern three different main types: perturbation methods, variational methods and asymptotic methods. The topic of the present book falls into this last category.

The merit of an asymptotic treatment is that it not only provides a convenient means for approximate calculation of details, but often it also has a fundamental physical significance. In our present context, the asymptotic treatment of the Schrödinger equation yields in its zeroth order the classical Hamilton–Jacobi equation, in its first order it yields the semi-classical approximation and in its higher orders it yields, in many important situations, a full account of quantal details, the physical quantities being sometimes obtained with an accuracy hardly obtainable by any other means.

The phase-integral method for the solution of physical problems used in this book is the result of a systematic work by the present authors and their coworkers over about four decades. The book is written with physical needs in mind. It contains three chapters. In Chapter 1, we give a historical survey. The purpose of Chapter 2 is to introduce the reader to the phase-integral method. The form of the wave function is discussed in Section 2.1, and the phase-integral approximation generated from an unspecified base function is presented in Section 2.2. In Section 2.3 the F-matrix method for the solution of connection problems is explained, and in Section 2.4 the parameterization of the F-matrix on opposite sides of a turning point is obtained. A thorough treatment of transmission through a real, single-hump potential barrier is given in Section 2.5, while transmission through a complex barrier is postponed to Section 3.44.

In Chapter 3 we present phase-integral solutions of quantal problems of various degrees of difficulty. Part of the material presented remedies unsatisfactory treatments by means of the so-called WKB approximation, which are often repeated in the literature. In general, our treatment is not restricted to the first order of the phase-integral approximation, and, due to the use of a convenient shorthand notation for certain contour integrals on a Riemann sheet, the formulas for an arbitrary-order phase-integral approximation do not have a more complicated appearance than those of the first-order approximation.

We would like to mention a few particular features of this book. The parameterization of the F-matrix connecting points on opposite sides of a turning point is presented in Section 2.4 for the first time, and hence the resulting formulas given there for the wave function on opposite sides of the turning point are new. They make it possible to obtain the formulas derived in Sections 3.21 and 3.24 and hence to solve, in the simple way shown in Section 3.29, the problem concerning the shifts of the energy levels due to compression of an atom, which previously was a very complicated problem. Several of the formulas for a real potential barrier in Section 2.5 and the connection formula for a complex potential barrier in Section 3.44 are also new. Formulas in the above-mentioned sections make it possible to obtain very precise formulas (3.22.10a, b) for the phase of the wave function with a given logarithmic derivative at the centre of a mirror symmetric potential barrier. The one-directional nature of the connection formulas associated with a turning point is discussed in detail in some of the problems in Chapter 3. Further evidence of their one-directional nature can be found in Fröman and Fröman (1998). The reason for the strange fact that one sometimes obtains a correct result, when one disregards the one-directional nature of the connection formulas is explained in Section 3.30. Examples of situations in which one obtains an erroneous result by using a connection formula when the conditions for its validity are not fulfilled (e.g., in the forbidden direction) are given in Sections 3.10, 3.13, 3.19, 3.20 and 3.45.

We remark that the notation used in this book differs from the notation in the original papers published up to the beginning of the 1980s in the respect that $Q^2(z)$, $Q_{\text{mod}}(z)$ and σ in those papers correspond in later publications and in the present book to $R(z)$, $Q(z)$ and $-\tilde{\phi}/2$, respectively.

In a collaboration several years ago A. Skorupski and P. O. Fröman invented the compound symbols and the 'traffic rules' presented in Section 2.3, which were found to be very useful in applications of the phase-integral method.

We would like to thank Professor Ulf Uhlhorn for valuable suggestions in the course of the work for this book and for great help with the construction of the figures. We are indebted to Dr Kilian Larsson for the calculation of the numerical results in Tables 3.16.1 and 3.16.2. We thank the Director of the Brera Observatory in Milan, Professor Guido Chincarini, for permission to publish Carlini's portrait in this book, and Dr Letizia Buffoni for the electronic transmission to us of a picture of the portrait. The help given by Sverker Nilsson, head of the Department of Print and Media at Uppsala University, in converting the manuscript and copying it onto a CD is gratefully acknowledged. We also thank Tamsin van Essen, Carol Miller and Maureen Storey at Cambridge University Press, who have been unfailingly helpful in all matters related to the publication of this book.

Nanny Fröman
Per Olof Fröman

1

Historical survey[†]

The mathematical approximation method which, since the breakthrough of quantum mechanics, has usually been called the WKB method, has in reality been known for a very long time. The method describes various kinds of wave motion in an inhomogeneous medium, where the properties change only slightly over one wavelength, and it also provides the connection between classical mechanics and quantum mechanics. To a surprisingly large extent it can already be found in an investigation by Carlini (1817) on the motion of a planet in an unperturbed elliptic orbit. After that the method was independently developed and used by many people. The important connection formulas were, however, missing, until Rayleigh (1912) very implicitly and Gans (1915) somewhat more explicitly derived one of them, which was later rediscovered independently by Jeffreys (1925), who also derived another connection formula (although not in quite the correct form), and by Kramers (1926).

1.1 Development from 1817 to 1926[‡]

1.1.1 Carlini's pioneering work

At the beginning of the nineteenth century Carlini (1817) (Fig. 1.1.1) treated an important problem in celestial mechanics. He considered the motion of a planet in an elliptic orbit around the sun, with the perturbations from all other gravitating bodies neglected. Using a polar coordinate system in the plane of the planetary motion, with the origin at the sun, one can express the polar angle as $2\pi t/T$ plus an infinite series containing sines of integer multiples of $2\pi t/T$, where t is the

[†] As a complement to our presentation of the historical development we refer the reader to McHugh (1971) and Schlissel (1977).
[‡] Section 1.1 is a somewhat revised version of the article by Fröman and Fröman (1985a) 'On the history of the so-called WKB-method from 1817 to 1926' in *Proceedings of the Niels Bohr Centennial Conference*, Copenhagen, March 25–28, 1985.

Figure 1.1.1 Francesco Carlini was born on 7 January 1783 and in 1799 became a student at the Brera Observatory in Milan. In 1832 he became head of the observatory and worked there until his death on 29 August 1862. This portrait is in the Brera Observatory in Milan.

time counted from a perihelion passage, i.e., from a moment when the planet is closest to the sun, and T is the time for one revolution of the planet in its orbit. The problem treated by Carlini was to determine the asymptotic behaviour of the coefficients of the sines in this series for large values of the summation index. In his treatment of this problem Carlini had to investigate a function s of a variable x. This function, which Carlini defined by a power series in x, is proportional to the function that is now called a Bessel function of the first kind, with the index p and the argument proportional to px. Carlini, who needed a useful approximate formula for this function when its argument is smaller than its order p, which tends to infinity, showed that $s(x)$ satisfies a linear, second-order differential equation containing the large parameter p. In this differential equation Carlini introduced a new dependent variable y by putting

$$s = \exp\left(\frac{1}{2}p\int^{x} y\, dx\right). \tag{1.1.1}$$

Then he expanded the function y in inverse powers of p. When Carlini introduced this expansion into the differential equation for y and identified terms containing the same power of $1/p$, he obtained recursive formulas which give what is now usually called the WKB approximation, with higher-order terms included, for the solution of the differential equation satisfied by $s(x)$. In explicit form he gave essentially the second-order WKB approximation for the solution in a classically forbidden region (in the language of quantum mechanics). If we express Carlini's result for the function $s(x)$ in terms of the Bessel function $J_p(\xi)$, where ξ is proportional to px, we obtain

$$
\begin{aligned}
J_p(\xi) = \frac{\xi^p}{2^p p!} \exp\Bigg\{ & p\left[\left(1 - \frac{\xi^2}{p^2}\right)^{1/2} - 1 - \ln \frac{1 + (1 - \xi^2/p^2)^{1/2}}{2}\right] \\
& - \frac{1}{2}\ln\left(1 - \frac{\xi^2}{p^2}\right)^{1/2} + \frac{1}{p}\left[\frac{1}{12} + \frac{1}{8(1 - \xi^2/p^2)^{1/2}}\right] \\
& - \frac{5}{24(1 - \xi^2/p^2)^{3/2}} + \cdots \Bigg\},
\end{aligned}
\tag{1.1.2}
$$

where $0 \le \xi < p$, ξ is proportional to p, and $p(>0)$ is large. Formula (1.1.2) is essentially equivalent to the next-to-lowest order of the asymptotic formula, derived almost a century later by Debye (1909), for the Bessel function $J_p(\xi)$ when $0 \le \xi < p$, ξ is proportional to p, and $p \to \infty$. We also remark that if in essential respects one follows Carlini's procedure to derive (1.1.2), but uses modern developments of the phase-integral technique, one can in a simple way obtain asymptotic formulas (P. O. Fröman, Karlsson and Yngve 1986), which are essentially equivalent to those derived by Debye with the aid of the more complicated method of steepest descents.

Using the language of quantum mechanics, one can say that in the part of the work by Carlini (1817) that has been described above, Carlini obtained an approximate expression for the solution of the radial Schrödinger equation in the classically forbidden region, in the absence of a *physical* potential. Because of the way in which the large parameter p appears in the differential equation for the function $s(x)$, Carlini's solution remains valid as $x \to 0$ in any order of approximation. Carlini thus automatically achieved in any order of approximation the result that Kramers (1926) achieved in the first-order WKB approximation by empirically replacing $l(l + 1)$ by $(l + 1/2)^2$, where l is the orbital angular momentum quantum number.

1.1.2 The work by Liouville and Green

In connection with a heat conduction problem, Liouville (1837) treated an ordinary, linear, second-order differential equation which he transformed into a differential equation of the Schrödinger type. Then he arrived at what one in quantal language now usually calls the first-order WKB approximation in a classically allowed region.

Green (1837) considered the motion of waves in a non-elastic fluid confined in a canal with infinite extension in the x-direction and with small breadth and depth, both of which may vary slowly in an unspecified way. The problem is described by a partial differential equation which is of second order with respect to both the coordinate x and the time t. Green obtained an approximate solution which, for the particular case in which its time dependence is described by a sine or cosine function, reduces essentially to the first-order WKB approximation in a classically allowed region.

As regards the work so far mentioned, it should be noted that Carlini did not treat a problem concerning wave motion, but one concerning planetary motion, and that he considered (in quantal language) a classically forbidden region. Liouville treated a problem concerning heat conduction, and Green treated a problem concerning waves in a fluid. In quantal language, both of these authors considered a classically allowed region.

1.1.3 Jacobi's contribution towards making Carlini's work known

The famous astronomer Encke, after whom a comet is named, drew Jacobi's attention to the work by Carlini (1817), and Jacobi (1849) published a paper concerning improvements and corrections to Carlini's work. In this paper Jacobi characterized Carlini's work as excellent and instructive, and he considered the problem treated in the main part of Carlini's publication as one of the most difficult problems of its class. Although Jacobi pointed out and corrected mistakes made by Carlini, he also pointed out that all the essential difficulties in the solution of the problem had been vanquished by Carlini (1817), and that Carlini's final result would have been correct if he had not made trivial mistakes in his calculations.

In 1850 Jacobi published a translation from Italian into German, with critical comments and extensions, of Carlini's investigation (Carlini 1850). In this publication Jacobi again emphasized that, although the work by Carlini (1817) contains many mistakes, and the final results are incorrect, this work, because of the method used there and the boldness of its composition, is still indisputably one of the most important works concerning the determination of the values of functions of large numbers. More than three decades after the original publication of Carlini's work, Jacobi thus considered it highly desirable to republish it with the necessary improvements and extensions included.

1.1.4 Scheibner's alternative to Carlini's treatment of planetary motion

The problem in celestial mechanics, which Carlini had treated by starting from a formula given by Lagrange, was later solved more generally and in much simpler ways by Scheibner (1856a,b), who attacked the problem from quite different starting points. In his first paper Scheibner (1856a) used a peculiar and very general method, which recommends itself by its brevity and ease of calculation. In his second paper Scheibner (1856b) used Cauchy's powerful theory of complex integration. As an indication of the importance of Scheibner's papers we mention that, almost a quarter of a century after they had first been published, the first paper (Scheibner 1856a), originally written in English, was republished in German translation (Scheibner 1880a), and the second paper (Scheibner 1856b), originally written in German, was republished in abbreviated form (Scheibner 1880b). Scheibner solved the actual problem in celestial mechanics much more simply and more satisfactorily than Carlini, but the more complicated investigation by Carlini yielded the very fundamental result that is now usually called the WKB approximation of arbitrary order. We mention Scheibner's work only to demonstrate the continued interest in the problem concerning planetary motion initiated by Carlini. The methods used by Scheibner are otherwise not related to the history of the so-called WKB method.

1.1.5 Publications 1895–1912

In his well-known book on hydrodynamics, Lamb (1895) treated (on pp. 291–6) the propagation of waves in a canal of gradually varying section on the basis of the investigation by Green (1837). Apparently unaware of the results obtained earlier by other authors, de Sparre (1898) derived essentially what is now called the second-order WKB approximation for a second-order differential equation. From a purely mathematical point of view, Horn (1899a,b) considered, for real values of the independent variable, the asymptotic solution of a linear, second-order differential equation containing a large parameter. Schlesinger (1906, 1907) generalized Horn's mathematical investigations by treating, for complex values of the independent variable, a linear system of first-order differential equations containing a large parameter. Referring to the method used by Green (1837), Birkhoff (1908) continued Horn's and Schlesinger's work by investigating mathematically the asymptotic character of the solutions of certain arbitrary-order linear differential equations containing a large parameter. With practical problems in mind, Blumenthal (1912) considered the asymptotic solution of a linear, second-order differential equation containing a large parameter. In a different way than Horn, he proved the existence of asymptotic approximations and obtained explicit estimates for their accuracy. Finally he applied his results to the differential equation for the spherical harmonics.

1.1.6 First traces of a connection formula

In a paper concerning the propagation of waves through a stratified medium, Rayleigh (1912) treated the one-dimensional time-independent wave equation by writing the solution as an amplitude times a phase factor. He found the exact relation between amplitude and phase (his eq. (73)), but he did not point out the great importance of this relation, which since 1930 has been used with great success by several authors for the numerical solution of differential equations of the Schrödinger type. Rayleigh obtained what is now generally known as the first-order WKB approximation in a classically allowed region. By pursuing the approximations he also obtained the next correction to the amplitude (in approximate form) and to the phase. Considering then the case of total reflection of waves due to a turning point, Rayleigh introduced into the wave equation a linear approximation in a certain region around the turning point and was thus able to obtain an approximate solution expressed as an Airy function in that region. When he used asymptotic approximations for the Airy function, he obtained a result that is closely related to a connection formula for the WKB approximation.

On the basis of Maxwell's electromagnetic theory, Gans (1915) treated the propagation of light in an inhomogeneous medium, where the index of refraction varies slowly and depends only on one cartesian coordinate. He obtained the first-order WKB approximation for the solution of the one-dimensional wave equation. When considering total reflection, Gans approximated the coefficient function in the differential equation by a linear function of the above-mentioned cartesian coordinate in the region around the turning point that gives rise to the total reflection. He was thus able to express the solution of the wave equation on each side of the turning point approximately in terms of Hankel functions of the order 1/3. Matching these approximate solutions at the turning point, and using asymptotic approximations for the Hankel functions on both sides of the turning point, Gans obtained a result (see eqs. (69) and p. 726 in his paper) which, although not in quite explicit form, is equivalent to the connection formula for the first-order WKB approximation that starts from the exponentially small wave function in the region into which the light penetrates only as an evanescent wave. This is, in somewhat more explicit form, the connection formula that Rayleigh (1912) had obtained.

1.1.7 Publications 1915–1921

In a paper dealing with certain hypotheses as to the internal structure of the earth and the moon, Jeffreys (1915) obtained (on pp. 211–213) essentially the first-order WKB approximation for the solution of a linear, second-order differential equation.

In an investigation concerning the aerodynamics of a spinning shell, Fowler, Gallop, Lock and Richmond (1921) treated a system of two coupled, ordinary, linear differential equations containing a large parameter, one of the equations being inhomogeneous and of the second order, the other being homogeneous and of the first order. Referring to the papers by de Sparre (1898), Horn (1899a,b), Schlesinger (1906, 1907) and Birkhoff (1908), Fowler *et al.* (1921) investigated the asymptotic expansion of the solution of the above-mentioned system of differential equations for large values of the parameter. In connection with this problem, the authors considered in particular a homogeneous, linear differential equation of the second order which they solved by writing the solution as the product of an amplitude and a phase factor. Finding the exact relation between amplitude and phase, they expressed the phase in terms of the amplitude which they obtained as an asymptotic expansion in inverse powers of the square of the large parameter. The authors made the important remark that by separating the solution correctly into the product of an amplitude and a phase factor they gained the advantage over other methods that they obtained in one step a solution with the error inversely proportional to the square of the large parameter, whereas this requires two steps in the usual procedures.

1.1.8 Both connection formulas are derived in explicit form

Referring to the above-mentioned work by Green, Lamb, Horn, Jeffreys and Fowler *et al.*, Jeffreys (1925) derived what is now usually called the WKB approximation for the solution of an ordinary, homogeneous, linear differential equation of the second order. In the region of a turning point Jeffreys, like Rayleigh and Gans, introduced a linear approximation into the differential equation and was thus able to express the solution there approximately in terms of Bessel functions of the order 1/3. Using asymptotic approximations for these functions, Jeffreys obtained the previously mentioned connection formula and another connection formula which was, however, not given in quite the correct form. The question of the one-directional nature of the connection formulas was not clarified until later.

1.1.9 The method is rediscovered in quantum mechanics

Brillouin (1926a,b) established, for a system of particles, the connection between the Schrödinger equation of quantum mechanics and the Hamilton–Jacobi equation of classical mechanics, while for the radial Schrödinger equation Wentzel (1926) and Kramers (1926), without knowing the work by previous authors, arrived at part of the results obtained in the course of the development described above. Kramers also pointed out that in the first order of the approximation it is sometimes convenient to replace $l(l + 1)$ by $(l + 1/2)^2$, where l is the orbital angular momentum quantum number. These results turned out to be extremely useful in applications of the new quantum theory and became known by the name of the WKB method. However, Brillouin, Wentzel and Kramers contributed hardly anything new to the mathematical approximation method that had already been found by previous authors, as described in this short historical review. Briefly speaking one can say that the so-called WKB method consists of the use of Carlini's approximation, which he derived in arbitrary-order approximation, and Jeffreys's connection formulas, which he derived in the first-order approximation. For the radial Schrödinger equation it also involves Kramers's modification of the first-order approximation by the replacement of $l(l + 1)$ by $(l + 1/2)^2$.

Since the publication of the papers by Brillouin (1926a,b), Wentzel (1926) and Kramers (1926) the method has been called the WKB method by most writers in theoretical physics, though this is not a very appropriate name. There are, however, some authors who have used more suitable names. It has thus been called the asymptotic approximation method by B. Swirles Jeffreys and the Liouville–Green method by H. Jeffreys and Olver and also by some other authors. Referring to the historical development described here we find it most natural that the so-called WKB approximation of arbitrary order (in which the whole asymptotic expansion is placed in the exponent of the exponential function, as one did for more than three decades after the publication of the papers by Brillouin, Wentzel and Kramers) should be called the Carlini approximation, and that the usual connection formulas for the first order of that approximation (associated with a well-isolated turning point) should be called Jeffreys' connection formulas. When the present authors expressed this opinion to Professor Clifford Truesdell some years ago, he replied that his teacher, Professor Bateman, famous for the California Institute of Technology Bateman Manuscript Project (1953, 1954, 1955), always used the name Carlini approximation instead of WKB approximation. In the present book we shall sometimes use the name Carlini approximation or Carlini (JWKB) approximation, in order to remind the reader of the origin of the approximation. However, we shall also use the name JWKB approximation or WKB approximation.

1.2 Development after 1926

Though the formulas of the WKB method became known to physicists in the 1920s, there were still a great number of questions to be answered concerning their accuracy, their range of applicability and, especially, the properties of the connection formulas. The problems were treated with varying rigour by many people.

Zwaan (1929) treated the connection problem in a new way. His idea was to allow the independent variable in the differential equation to take complex values and to derive a connection formula by tracing the solution in the complex plane around the critical point. To quote Birkhoff (1933): 'Zwaan's treatment is extremely suggestive, although lacking in essential respects'. An attempt to put Zwaan's method on a rigorous basis was made in a well-known paper by Kemble (1935); see also Kemble (1937). He transformed the original linear differential equation of the second order in a very convenient way to a system of two linear differential equations of the first order, which he integrated by an approximate method. Furry

(1947) used Zwaan's (1929) approach for the treatment of Stokes's phenomenon and derived the connection formulas by a new method. He also calculated the normalization integral for a bound state without assuming the quantum number to be large.

The question concerning the one-directional nature of the connection formulas associated with a turning point led to a well-known debate between Langer (1934) and Jeffreys (1956). Their dispute was, however, due to misunderstandings. In fact, they both asserted that the connection formulas are one-directional. Discussions concerning the connection formulas were later taken up by Fröman and Fröman (1965, 1998), Dingle (1965, 1973), Berry and Mount (1972) and Silverstone (1985). The confusion concerning the properties of the connection formulas derives from a lack of rigour in the very formulation of the connection problem.

Heading (1962) published the first book that was completely devoted to the WKB method. In the preface he wrote '. . . it is surprising that the development of this technique over the last fifty years has been the occasion of so much error, criticism and dispute. Moreover, the treatment of the subject in the literature ranges from the ridiculously simple void of all rigour to the most sophisticated, the former hardly deserving mention and the latter not forming part of what is commonly known as the W. K. B. J. method' and somewhat later '. . . its name is known to many but its actual technique is known to but few.' Heading aimed at presenting this technique in his book. However, unsatisfactory elements remained in the method.

In spite of the abundant literature, at the beginning of the 1960s there still was not a convenient method for obtaining definite limits of error in more general cases, and Heading (1962), see his p. 59, expressed the opinion that 'This vagueness must be accepted as one of the inherent weaknesses of the phase-integral method'. The problem of obtaining limits of error is not only of academic or mathematical interest but is especially important in some physical applications, where lack of rigour may imply that one cannot have confidence in the results.

Using as a starting point the ideas introduced by Zwaan (1929) and Kemble (1935), the present authors found that the system of two first-order differential equations introduced by Kemble has an exact solution in the form of fairly simple convergent series. Exploiting this result, we developed in 1960–4 a new, rigorous method for handling the connection problems for the so-called WKB approximation of the first order and modifications of it. The study of connection problems was thereby transformed to the study of a certain matrix, the F-matrix, the elements of which were given by convergent series. This method, which is also powerful in intricate and complicated applications, was presented in a monograph (Fröman and Fröman 1965). N. Fröman (1966b) generalized the treatment there and showed that one could start with an ordinary differential equation of arbitrary order, assume a set of functions representing approximate solutions, and derive an exact

solution of the original differential equation that could be used for solving connection problems. An advantage of the approach in these publications is that one
works with an exact solution and makes all approximations in the final formulas, yet
has a close contact with the approximate formulas in all steps of the calculations.
Fröman and Fröman (1965) provided a sound basis for handling the connection
problems of the first-order WKB approximation and led to satisfactory estimates of
the accuracy of that approximation. Soon after its publication, Olver (1965a,b) published related estimates, which he had derived quite independently using another
approach.

In the 1960s considerable interest was focused on the study of modifications
of the WKB approximation in higher orders with the purpose of obtaining an
approximate solution of the radial Schrödinger equation that remains valid at the
origin. Choi and Ross (1962) and Krieger and Rosenzweig (1967) gave important
contributions to the solution of this problem, but they did not give an explicit, simple
asymptotic expression of arbitrary order for the wave function. Such an asymptotic
expression was given by Fröman and Fröman (1974a,b) with the derivation of
the phase-integral approximation generated from an unspecified base function.
This new approximation, which is related to the WKB approximation, although
with important advantageous differences, is described briefly in Section 2.2 of the
present book and in detail in Chapter 1 of Fröman and Fröman (1996).

As already mentioned, Jeffreys (1925) derived for the first-order WKB approximation connection formulas associated with a turning point, but to the authors'
knowledge no one derived corresponding higher-order connection formulas.
N. Fröman (1970) derived for an arbitrary-order phase-integral approximation connection formulas associated with a turning point, which later turned out to be valid
also for the phase-integral approximation generated from an unspecified base function. These general connection formulas contain the phase integrand $q(z)$ and the
phase integral $w(z) = \int_t^z q(z)dz$, where t is the turning point, in the same way
for any order of the phase-integral approximation, which was not at all clear until
the arbitrary-order connection formulas in question had actually been derived; the
connection formula for a real single-hump potential barrier, given by (3.45.5a,b),
(3.45.8), (3.45.9a) and (3.45.13), is, for instance, quite different in different orders
of approximation.

We also mention the development of phase-integral formulas not involving wave
functions for normalization integrals (Furry 1947; de Alfaro and Regge 1965
pp. 64–5; Yngve 1972; P. O. Fröman 1974), expectation values (Delves 1963;
Dagens 1969; Siebert and Krieger 1970; N. Fröman 1974) and matrix elements
(Fröman and Fröman 1977; Fröman, Fröman and Karlsson 1979; P. O. Fröman
2000). The analytic matrix element formula for unbound states given by Fröman,
Fröman and Karlsson (1979) yielded results of much higher accuracy than one could

obtain numerically; see Section 3.43. Other situations in which phase-integral results were at least as accurate as results obtained numerically occurred for the Stark effect in some levels of a hydrogen atom; see Section 3.48.

A supplementary quantity (related to $\tilde{\phi}$ in the present book) that appears in the connection formula for a real potential barrier was obtained in the first-order approximation by Ford, Hill, Wakano and Wheeler (1959) and in the first-, third- and fifth-order approximations by Fröman, Fröman, Myhrman and Paulsson (1972). The corresponding supplementary quantity ϕ for a complex barrier was obtained up to the thirteenth order of approximation by Fröman, Fröman and Lundborg (1996).

Fröman and Fröman (1996) adapted the comparison equation technique, devised chiefly by Cherry (1950) and Erdélyi (1956, 1960), to the phase-integral technique in order to be able to calculate analytic expressions for the supplementary quantities (F-matrix elements, suitably called Stokes constants) needed in order to master connection problems when transition points approach each other. We performed the rather lengthy calculations up to the fifth order of the phase-integral approximation generated from an unspecified base function. The formulas thus obtained can readily be particularized to specific situations. Various applications were treated in adjoined papers in Fröman and Fröman (1996) in order to illustrate the accuracy that can be achieved in physical applications by means of the phase-integral method with the Stokes constants obtained according to the comparison equation technique.

Three-dimensional phase-integral investigations have been published by Glaser and Braun (1954, 1955) for a classically allowed region and by P. O. Fröman (1957) for a classically forbidden region. Work on multi-dimensional phase integrals goes back to Maslov's appendix in the Russian translation of Heading (1962). Further work along these lines is described in Maslov and Fedoriuk (1981), but we shall not discuss this approach here. Nor do we consider the phase-integral treatment of coupled differential equations. In the present book we thus restrict ourselves to considering problems that can be reduced to the treatment of a one-dimensional differential equation of the Schrödinger type.

2

Description of the phase-integral method

2.1 Form of the wave function and the q-equation

A general one-dimensional differential equation of the Schrödinger type can be written as

$$d^2\psi/dz^2 + R(z)\psi = 0, \qquad (2.1.1)$$

where $R(z)$ is assumed to be a single-valued analytic function of the complex variable z in a certain region of the complex z-plane. In the case of the Schrödinger equation we have with obvious notation $R(z) = 2m[E - V(z)]/\hbar^2$, where $V(z)$ may be the actual physical potential or an effective potential. Thus, if we are concerned with the radial Schrödinger equation, $V(z)$ also includes the centrifugal potential.

It is very natural to look for two linearly independent solutions of (2.1.1) of the form

$$\psi = A(z)\exp[\pm i w(z)], \qquad (2.1.2)$$

where $A(z)$ and $w(z)$ are functions, the properties of which are to be determined. Without introducing any reality conditions on the functions $R(z)$, $A(z)$ and $w(z)$, we note that the Wronskian of two linearly independent solutions of (2.1.1) is a constant $\neq 0$, and that the Wronskian of the two functions (2.1.2) is equal to $-2i A^2(z)dw(z)/dz$ and is thus constant only if $A = \text{const} \times (dw/dz)^{-1/2}$. Writing $q(z)$ instead of $dw(z)/dz$, we thus see that the requirement that the Wronskian of the two functions (2.1.2) be constant implies that these functions (except for arbitrary constant factors) can be written in the form

$$\psi = q^{-1/2}(z)\exp[\pm i w(z)], \qquad (2.1.3a)$$

$$w(z) = \int^z q(z)dz. \qquad (2.1.3b)$$

The function $w(z)$ is the phase integral, and therefore we call $q(z)$ the phase integrand. Although $q(z)$ and $w(z)$ are in general complex, and may also be so on the real z-axis, one may in (2.1.3a) call $q^{-1/2}(z)$ the complex 'amplitude' and $\pm w(z)$ the complex 'phase'. The use of solutions of the form (2.1.3a,b), possessing the property of a constant Wronskian, is of decisive importance in the phase-integral method. The two functions (2.1.3a,b) are linearly independent, since their Wronskian (being equal to $-2i$) is different from zero. It should be emphasized that the *exact* phase integrand $q(z)$ introduced here is sometimes not even approximately the same as the *asymptotic* phase integrand $q(z)$ that appears in the phase-integral approximation generated from an unspecified base function; see Section 2.2. In passing we also draw attention to the fact that if $R(z)$ is real on the real z-axis, every solution of (2.1.1) that does not have a constant phase on this axis, i.e., that is not equal there to a real function times a possibly complex constant, can (except for a constant factor) be written in the form (2.1.3a,b) with $q(z)$ real on the real z-axis.

Inserting (2.1.3a,b) into (2.1.1), we find that the exact phase-integrand $q(z)$ must be a solution of the differential equation

$$q^{-3/2}d^2q^{-1/2}/dz^2 + R(z)/q^2 - 1 = 0, \tag{2.1.4}$$

which we call the q-equation. For any solution $q(z)$ of (2.1.4), the functions (2.1.3a,b) are linearly independent solutions of (2.1.1).

The results obtained in this section form the starting point for the derivation in Section 2.2 of the phase-integral approximation generated from an unspecified base function.

2.2 Phase-integral approximation generated from an unspecified base function

One might at first sight think that a natural way of improving the accuracy of the first-order WKB approximation would be to use a higher-order WKB approximation. However, the higher-order WKB functions are not of the simple form (2.1.3a,b), and their Wronskian is therefore not constant; see Section 1.2 of Fröman and Fröman (1996). Because of this fact difficulties appear when one tries to solve connection problems for the higher orders of the WKB approximation. Therefore in this section we shall derive a phase-integral approximation of the simple form (2.1.3a,b).

The correct relation between 'phase' and 'amplitude', shown in (2.1.3a,b), was used in the 1930s for the *numerical solution* of the Schrödinger equation, but there are very few publications before 1966 in which there are traces of *higher-order asymptotic approximations* with the correct relation between 'phase' and 'amplitude'. If actually pursued to higher order, the procedure used by Messiah (1959, Chapter VI, Section 7) would have yielded such a kind of higher-order asymptotic approximation instead of the WKB approximation of higher order. However, the distinction between these two kinds of approximations does not appear in Messiah's treatment, since he restricts himself to deriving the first-order approximation only. Broer (1963) used the same procedure as Messiah and pursued it by deriving the first two higher-order approximations, but in his terminology he did not distinguish the resulting higher-order approximations from the higher-order WKB approximations. This new kind of higher-order asymptotic approximation also appears implicitly in a paper by Bertocchi, Fubini and Furlan (1965), but the advantage of using these approximations for solving the connection problems is neither indicated nor made use of; see the paper by Dammert and P. O. Fröman (1980) in which the important difference between the two types of approximations was emphasized. It was only in 1966 that asymptotic approximations of arbitrary order, with the correct relation between 'phase' and 'amplitude', shown in (2.1.3a,b), were derived in a systematic way by N. Fröman (1966d). Using the recurrence formula for the WKB series, she showed that the usual WKB series, with infinitely many terms retained, can be transformed so that the sum of the terms of odd order in the expansion parameter can be simply expressed in terms of the sum of the terms of even order, and she thus obtained a formal solution of the time-independent Schrödinger equation containing only the even-order terms. This transformation is exact when the whole series is retained. If the series is truncated, a new approximation is obtained, which in higher order differs from the WKB approximation of corresponding order; see Dammert and P. O. Fröman (1980) and Chapter 1 in Fröman and Fröman (1996). The distinguishing property is that the new approximation displays, also in higher order, the simple relation, shown in (2.1.3a,b), between the pre-exponential factor and the integrand of the integral in the exponent. An important consequence of this simple analytical form is that the Wronskian of two approximate solutions is constant, as it should be, since the Wronskian of two exact solutions is constant, and the F-matrix method for mastering connection problems (Fröman and Fröman 1965) could be applied to the new approximation after only a minor generalization. However, this approximation is not flexible enough for the treatment of important, commonly appearing problems, for instance in connection with the radial Schrödinger equation when the effective potential has a first- or second-order pole at the origin. An asymptotic approximation free from this deficiency was derived by Fröman and Fröman (1974a,b). The flexibility is achieved by the introduction of an *a priori* unspecified *base function* $Q(z)$, from which the approximation is generated. For a detailed description of the phase-integral approximation generated from an unspecified base function we refer the reader to Chapter 1 in Fröman and Fröman (1996), and for the advantages of this approximation compared with the

WKB approximation we refer the reader to Dammert and P. O. Fröman (1980) and also to Fulling (1983), who mentions an intended application to general-relativistic quantum field theory. A brief description of the former approximation is given below.

We assume that in some way or another we have determined a function $Q(z)$ that is an approximate solution of the q-equation (2.1.4). This implies that the quantity ε_0 defined by the left-hand side of (2.1.4) with q replaced by $Q(z)$, i.e.,

$$\varepsilon_0 = Q^{-3/2}(z)d^2Q^{-1/2}(z)/dz^2 + R(z)/Q^2(z) - 1 \qquad (2.2.1)$$

$$= \frac{1}{16Q^6}\left[5\left(\frac{dQ^2}{dz}\right)^2 - 4Q^2\frac{d^2Q^2}{dz^2}\right] + \frac{R-Q^2}{Q^2}, \qquad (2.2.1')$$

be small compared with unity. We take this smallness explicitly into account by considering ε_0 to be proportional to λ^2, where λ is a 'small' bookkeeping parameter. This is attained if $Q(z)$ is proportional to $1/\lambda$ and $R(z) - Q^2(z)$ is independent of λ, i.e., if $R(z)$ is replaced by $Q^2(z)/\lambda^2 + [R(z) - Q^2(z)]$. Therefore we consider instead of the original differential equation (2.1.1) the auxiliary differential equation

$$d^2\psi/dz^2 + \{Q^2(z)/\lambda^2 + [R(z) - Q^2(z)]\}\psi = 0, \qquad (2.2.2)$$

which goes over into (2.1.1) when $\lambda = 1$. We have thus introduced the 'small' parameter λ, which at the end is to be put equal to unity, and which only plays the role of a bookkeeping parameter. The old question concerning which parameter is actually to be considered as small in the Schrödinger equation can now be answered: There is from the beginning no quantity that can be used as such a 'small' parameter in every situation that occurs. The parameter has instead to be introduced in a way that is convenient for each particular problem. Often one can use Planck's constant h as such a parameter, but that cannot be done, for instance, in connection with the radial Schrödinger equation where the well-known replacement of $l(l + 1)$ by $(l + 1/2)^2$ in the first-order approximation is used in order that the approximate wave function be valid at the origin.

Inserting (2.1.3a,b) into (2.2.2), we obtain for $q(z)$ the differential equation

$$q^{+1/2}d^2q^{-1/2}/dz^2 - q^2 + Q^2(z)/\lambda^2 + R(z) - Q^2(z) = 0, \qquad (2.2.3)$$

which is called the auxiliary q-equation, and which goes over into the original q-equation (2.1.4) when one puts $\lambda = 1$. Replacing differentiation with respect to z by differentiation with respect to the variable

$$\zeta = \int^z Q(z)dz, \qquad (2.2.4)$$

we can write (2.2.3) in the form

$$1 - \left[\frac{q\lambda}{Q(z)}\right]^2 + \varepsilon_0\lambda^2 + \left[\frac{q\lambda}{Q(z)}\right]^{+1/2}\frac{d^2}{d\varsigma^2}\left[\frac{q\lambda}{Q(z)}\right]^{-1/2}\lambda^2 = 0, \quad (2.2.5)$$

where ε_0 is defined by (2.2.1). To obtain a formal solution of (2.2.5), we put

$$q\lambda/Q(z) = \sum_{n=0}^{\infty} Y_{2n}\lambda^{2n}, \quad (2.2.6)$$

where Y_0 is assumed to be different from zero, and Y_{2n} $(n = 0, 1, 2, \ldots)$ is independent of λ. Inserting the expansion (2.2.6) into (2.2.5), expanding the left-hand side in powers of λ, and putting the coefficient of each power of λ equal to zero, we get $Y_0 = \pm 1$ and a recurrence formula, from which one can successively obtain the functions $Y_2, Y_4, Y_6, \ldots$, each one of which can be expressed in terms of ε_0, and derivatives of ε_0 with respect to ς. Since we have introduced the double sign $\pm$ in (2.1.3a), it is no restriction to choose $Y_0 = 1$. The first few functions Y_{2n} are then

$$Y_0 = 1, \quad (2.2.7a)$$

$$Y_2 = \frac{1}{2}\varepsilon_0, \quad (2.2.7b)$$

$$Y_4 = -\frac{1}{8}\left(\varepsilon_0^2 + \frac{d^2\varepsilon_0}{d\zeta^2}\right), \quad (2.2.7c)$$

$$Y_6 = \frac{1}{32}\left[2\varepsilon_0^3 + 5\left(\frac{d\varepsilon_0}{d\zeta}\right)^2 + 6\varepsilon_0\frac{d^2\varepsilon_0}{d\zeta^2} + \frac{d^4\varepsilon_0}{d\zeta^4}\right]. \quad (2.2.7d)$$

The choice of the unspecified base function $Q(z)$ shows itself only in the expressions (2.2.1) and (2.2.4) for ε_0 and ς, respectively, as functions of $R(z)$ and $Q(z)$, while the expressions for the functions Y_{2n} in terms of ε_0 and derivatives of ε_0 with respect to ς do not depend explicitly on $R(z)$ and the choice of the base function $Q(z)$. The expressions for the functions Y_{2n} can therefore be determined once and for all, and explicit expressions up to Y_{20} were given by Campbell (1972). A number of general properties of Y_{2n} were found by Skorupski (1980, 1988).

It is often of great practical importance to note that one can write

$$Y_{2n} = Z_{2n} + dU_{2n}/d\zeta, \quad (2.2.8)$$

where the splitting of the right-hand side into two terms is uniquely determined by the requirement that in every term in the expression for Z_{2n} the indices ν of the quantities ε_ν be as small as possible. From (2.2.7a–d) one thus finds that

$$Z_0 = 1, \qquad\qquad\qquad (2.2.9a)$$

$$Z_2 = \frac{1}{2}\varepsilon_0, \qquad\qquad\qquad (2.2.9b)$$

$$Z_4 = -\frac{1}{8}\varepsilon_0^2, \qquad\qquad\qquad (2.2.9c)$$

$$Z_6 = \frac{1}{32}\left[2\varepsilon_0^3 - \left(\frac{d\varepsilon_0}{d\zeta}\right)^2\right], \qquad\qquad (2.2.9d)$$

and

$$U_0 = 0, \qquad\qquad\qquad (2.2.10a)$$
$$U_2 = 0, \qquad\qquad\qquad (2.2.10b)$$

$$U_4 = -\frac{1}{8}\frac{d\varepsilon_0}{d\zeta}, \qquad\qquad\qquad (2.2.10c)$$

$$U_6 = \frac{1}{32}\left[6\varepsilon_0\frac{d\varepsilon_0}{d\zeta} + \frac{d^3\varepsilon_0}{d\zeta^3}\right]. \qquad\qquad (2.2.10d)$$

Explicit expressions for Z_{2n} and U_{2n} up to $2n = 16$ were given by Campbell (1979). Truncating the infinite series in (2.2.6) at $n = N$ and putting $\lambda = 1$, we obtain

$$q(z) = Q(z)\sum_{n=0}^{N} Y_{2n}. \qquad\qquad (2.2.11)$$

Inserting (2.2.11) into (2.1.3a,b), we get the phase-integral approximation of the order $2N + 1$, generated from the base function $Q(z)$, as the approximate solution of the original differential equation (2.1.1). At zeros and simple poles of $Q^2(z)$ the functions $q(z)$ and $q^{1/2}(z)$ are in general singular and may have branch points, whereas the functions ε_0, Y_{2n} and $q(z)/Q(z)$ are single-valued, when $R(z)$ is single-valued. For $N > 0$ the function $q(z)$ has simple zeros in the neighbourhood of each transition zero (N. Fröman 1970). In the first order, the approximation now described is the same as the usual WKB approximation if $Q(z) = R^{1/2}(z)$, but in higher orders it differs in essential respects from the WKB approximation of corresponding order; see N. Fröman (1966d), Dammert and P. O. Fröman (1980), and Fröman and Fröman (1996).

It is a fundamental advantage of the phase-integral approximation described above over the WKB approximation in higher order that the former contains the unspecified base function $Q(z)$, which one can utilize in several ways. A criterion for the determination of the base function is that the function ε_0 be small compared with unity in the region of the complex z-plane relevant for the problem under consideration. However, this criterion does not determine the base function $Q(z)$ uniquely; it turns out that, within certain limits, the results are not very sensitive to the choice of $Q(z)$, when the approximation is used in higher orders. An inconvenient, but possible, choice of $Q(z)$ introduces an unnecessarily large error in the first-order approximation that is, however, in general already corrected in the third-order approximation. In many important cases the function $Q^2(z)$ can be chosen to be identical to $R(z)$. In other important cases, for instance when one wants to include the immediate neighbourhood of a first- or second-order pole of $R(z)$ in the region of validity of the phase-integral approximation, the function $Q^2(z)$ is chosen to be similar to $R(z)$ except in the neighbourhood of the pole.

The freedom in the choice of the base function $Q(z)$ will now be illuminated in a more concrete way. For a radial Schrödinger equation the usual choice of $Q^2(z)$ is

$$Q^2(z) = R(z) - \frac{1}{4z^2}. \qquad (2.2.12a)$$

However, the replacement of (2.2.12a) by

$$Q^2(z) = R(z) - \frac{1}{4z^2} - \frac{\text{const}}{z}, \qquad (2.2.12b)$$

where the coefficient of $1/z$ should be comparatively small, does not destroy the great accuracy of the results usually obtained with the phase-integral approximation in higher orders. There is thus a whole set of base functions that may be used, and there are various ways in which one can take advantage of this non-uniqueness to make the choice of the base function well adapted to the particular problem under consideration. For instance, by adapting the choice of $Q^2(z)$ to the analytical form of $R(z)$ one can sometimes achieve the result that the integral occurring in the phase-integral approximation can be evaluated analytically. To give an example, we assume that $R(z)$ contains only $\cot z$ but not z itself. In this case it is convenient to replace the choice (2.2.12b) by the choice

$$Q^2(z) = R(z) - \tfrac{1}{4}\cot^2 z - \text{const} \times \cot z. \qquad (2.2.12c)$$

By a convenient choice of $Q^2(z)$, for instance as concerns the unspecified coefficient in (2.2.12b) or (2.2.12c), one can try to make the first-order approximation as accurate as possible. Sometimes one can even attain the result that, for example, eigenvalues or phase shifts are obtained exactly for some particular parameter value in every order of the phase-integral approximation. By making this exactness fulfilled in the limit of a parameter value, for which the phase-integral result without this adaptation would deteriorate, one can actually extend the region of validity of the phase-integral treatment; see pp. 16–17 in Fröman, Fröman and Larsson (1994). When the differential equation contains one or more parameters, the accurate calculation of the wave function may require different choices of the base function $Q(z)$ for different ranges of the parameter values. To illustrate this fact we consider the radial Schrödinger equation. For sufficiently large values of the angular momentum quantum number l we obtain an accurate phase-integral approximation (which is also valid close to $z = 0$) if we choose $Q^2(z)$ according to (2.2.12a) or (2.2.12b). If the value of l is too small, the corresponding phase-integral approximation is not good. It can be considerably improved (except close to $z = 0$) when the absolute value of the coefficient of $1/z$ in $R(z)$ is positive and sufficiently large if instead one chooses

$$Q^2(z) = R(z) + \frac{l(l+1)}{z^2}. \tag{2.2.12d}$$

The corresponding phase-integral approximation is not valid close to $z = 0$, but the wave function that is regular and tends to z^{l+1} as $z \to 0$ can be obtained sufficiently far away from $z = 0$ by means of a particular connection formula; see Section 3.38 and Fröman and Fröman (1965), eq. (7.28).

When considering the integral of $q(z)$ along a closed contour Λ on which $Q(z)$ is single-valued, we obtain with the aid of (2.2.11), (2.2.8) and (2.2.4)

$$\int_\Lambda q(z)dz = \sum_{n=0}^{N} \int_\Lambda Y_{2n} Q(z)dz = \sum_{n=0}^{N} \int_\Lambda Z_{2n} Q(z)dz, \tag{2.2.13}$$

where the first few functions Z_{2n} are given by (2.2.9a–d).

When the first-order approximation is used, it is often convenient to choose the constant lower limit of integration in the definition (2.1.3b) of the phase integral $w(z)$ to be a zero or a first-order pole t of $Q^2(z)$. This is not possible, however, when a higher-order approximation is used, since the phase integral would then in general be divergent. If t is a transition point of odd order, it is therefore convenient to replace the original definition of $w(z)$ by the definition (N. Fröman 1966d, 1970)

$$w(z) = \frac{1}{2} \int_{\Gamma_t(z)} q(z)dz, \tag{2.2.14}$$

where $\Gamma_t(z)$ is a path of integration that starts at the point corresponding to z on a Riemann sheet adjacent to the complex z-plane under consideration, encircles t in the positive or in the negative sense and ends at z. It is immaterial for the value of the integral in (2.2.14) if the path of integration encircles t in the positive or in the negative sense, but the endpoint must be z. In the first-order approximation the expression on the right-hand side of (2.2.14) can be replaced by the usual integral of $Q(z)$ from t to z. However, even for the first-order approximation it may be preferable to replace the integration along the real axis by integration along a contour on a Riemann sheet, since then an accurate numerical determination of the position of t is not needed, and this makes it possible to calculate the integral more accurately.

It is very useful to introduce a short-hand notation for the contour integral on the right-hand side of (2.2.14). To do this, we write instead of (2.2.14)

$$w(z) = \int\limits_{(t)}^{z} q(z)dz, \qquad (2.2.15)$$

the integral in (2.2.15) being by definition equal to the integral in (2.2.14). In the first-order approximation one can replace (t) by t in (2.2.15) and thus get an ordinary integral from t to z instead of half of the integral along the contour $\Gamma_t(z)$.

We introduced the notation on the right-hand side of (2.2.15) under the assumption that t is an odd-order zero or pole of $Q^2(z)$, but now we extend (2.2.15) to cases in which t is not a transition point, and hence $q(z)$ is regular in the neighbourhood of t, by defining in these cases the integral in (2.2.15) to be just the integral in (2.1.3b) with the constant lower limit of integration equal to t. Thus the integral in (2.2.15) is now defined for all points t that are *not* transition points of *even* order. In a similar way one defines a short-hand notation for an integral in which the upper limit of integration is *not* a transition point of *even* order. When one has two transition points of odd order as limits of integration, one requires that the contours of integration pertaining to the upper and lower limits are encircled in the same direction, and the definition of the short-hand notation with both limits of integration within parentheses then implies that the integral is equal to half of the integral along a closed loop enclosing both transition points. With the notation on the right-hand side of (2.2.15) the formulas in an arbitrary-order approximation look almost like those in the first-order approximation, and one thus achieves a great formal and practical simplification of the work. The short-hand notation on the right-hand side of (2.2.15) for the integral on the right-hand side of (2.2.14) was introduced by Fröman, Fröman and Lundborg (1988a), pp. 160–1. It makes it possible to use, for an arbitrary order of the phase-integral approximation, a similar simple notation and almost the same simple language (although in a generalized sense) as for the first order of the phase-integral approximation. This really simplifies the treatment of concrete problems when an arbitrary order of the phase-integral approximation is used.

We have seen that solutions of the form (2.1.3a,b) are linearly independent and have a constant Wronskian, and that for the function $q(z)$ in (2.1.3a,b) one can obtain fairly simple asymptotic expressions with respect to a 'small' parameter. Thus one obtains the phase-integral approximation generated from an unspecified base function. However, if one looks for linearly independent asymptotic approximations of the Carlini (JWKB) form, i.e., solutions with the whole asymptotic series in the exponent, one is not able to obtain asymptotic approximations with a constant Wronskian, nor to express these solutions in terms of a single function. The Carlini approximation involves two functions and can be written as

$$\psi = \exp[u(z) \pm iw(z)], \tag{2.2.16}$$

where $u(z)$ and $w(z)$ are real functions, when z and $R(z)$ are real. The function $w(z)$ in (2.2.16) is the same as the function $w(z)$ in the phase-integral approximation generated from an unspecified base function with $Q^2(z) = R(z)$, but the function $u(z)$ is very complicated in the higher orders of the Carlini approximation; see eqs. (1.2.6a–c) in Fröman and Fröman (1996).

The rest of the present book is based on the use of the phase-integral approximation generated from an unspecified base function that has been presented in the this section.

2.3 *F*-matrix method

The requirement that $q(z)$ be an approximate solution of the q-equation (2.1.4) is sufficient for solving the *local* problem of finding approximate solutions of the differential equation (2.1.1). However, in physical applications it is usually necessary to know the behaviour of one and the same solution of a differential equation not only locally in the neighbourhood of a certain point, but over extended regions of the independent variable. One then encounters difficulties even if the condition $|\varepsilon_0(z)| \ll 1$, with $\varepsilon_0(z)$ defined by (2.2.1), is excellently fulfilled at every point in the region of interest. If we allow the independent variable to take values in the complex z-plane, the same linear combination may be used within a certain region of the complex z-plane, whereas in other regions different linear combinations have to be used to represent the same solution ψ of the differential equation. This is the well-known *Stokes phenomenon*. The problem of determining the linear combinations of asymptotic solutions that represent the same solution ψ in different regions is called the *connection* problem or the *global* problem. It is deeper than the local problem, and it is of vital importance. Thus, even if one knows a set of linearly independent functions that satisfy a given differential equation approximately, and even to a very high degree of accuracy, it may still be a difficult matter to use these solutions for solving problems encountered in theoretical physics. The boundary conditions imposed by the physical problem require a special solution to be represented approximately. However, it may occur that a linear combination of approximate solutions, satisfying a boundary condition, does not approximately represent the exact solution when one proceeds away from the boundary.

It is very important to formulate the connection problem in a precise way. Lack of clarity in that respect can give rise to misconceptions, which have not failed to show themselves in the literature. With regard to the needs for the solution of physical problems, the natural formulation, which we shall adopt, is the following: *Given a linear combination of phase-integral functions representing a certain solution in one part of the complex z-plane, we ask which linear combination is appropriate for approximating that solution in another part of the complex z-plane.* With this formulation, the relation (2.3.7) below displays the connection problem in a condensed mathematical form.

In this section we shall briefly present a rigorous method for mastering the Stokes phenomenon and hence for the solution of connection problems, the *F*-matrix method (Fröman and Fröman 1965), which constitutes an essential part of the phase-integral method. The *F*-matrix method was originally developed for mastering the connection problems for the first order of the WKB approximation and simple modifications of it, but, with the slight changes introduced in this section, it can be applied to an arbitrary order of the phase-integral approximation described in Section 2.2. In the presentation below the notation differs from that in Fröman and Fröman (1965) in the respect that the functions, which below are called $R(z)$ and $Q(z)$, were in Fröman and Fröman (1965) called $Q^2(z)$ and $q(z)$, respectively. Furthermore, our matrix $\mathbf{M}(z)$, defined by (2.3.5b) below, differs from the matrix $\mathbf{M}(w)$ defined by eq. (3.13) in Fröman and Fröman (1965), but is simply related to that matrix.

2.3.1 Exact solution expressed in terms of the F-matrix

We introduce the phase-integral functions $f_1(z)$ and $f_2(z)$, which are assumed to be approximate solutions of the original differential equation (2.1.1), according to the definitions (cf. (2.1.3a,b) and (2.2.15))

$$f_1(z) = q^{-1/2}(z) \exp[+i w(z)], \qquad (2.3.1a)$$

$$f_2(z) = q^{-1/2}(z) \exp[-i w(z)], \qquad (2.3.1b)$$

$$w(z) = \int_{(Z)}^{z} q(z)dz, \qquad (2.3.1c)$$

where $q(z)$ is an analytic function, and Z is an unspecified fixed point. We remark that the treatment in this section is valid even if $q(z)$ is not given by (2.2.1). The complex z-plane is assumed to be cut in such a way that $q^{-1/2}(z)$ and $w(z)$ are single-valued. The functions given by (2.3.1a–c) possess the same property as exact solutions of the differential (2.1.1) in the sense that their Wronskian is constant,

$$\begin{vmatrix} f_1(z) & f_2(z) \\ f_1'(z) & f_2'(z) \end{vmatrix} = f_1(z)f_2'(z) - f_2(z)f_1'(z) = -2i, \qquad (2.3.2)$$

and they satisfy one and the same differential equation, *viz*

$$d^2f_k/dz^2 + \left(q^2 - q^{1/2}\,d^2q^{-1/2}/dz^2\right)f_k = 0, \quad k = 1, 2, \qquad (2.3.3)$$

which becomes the same as (2.1.1) when the q-equation (2.1.4) is satisfied. The relations (2.3.2) and (2.3.3) are of crucial importance for the practical usefulness of the F-matrix method. They express distinctive properties of the higher-order phase-integral functions as compared to the higher-order WKB functions, which do not have a constant Wronskian and do not satisfy one and the same differential equation.

We introduce two functions, the a-coefficients $a_1(z)$ and $a_2(z)$, which are uniquely determined by the requirements that any exact solution ψ of the differential equation (2.1.1), together with its derivative, can be written exactly as

$$\psi = a_1(z)f_1(z) + a_2(z)f_2(z), \qquad (2.3.4a)$$

$$d\psi/dz = a_1(z)df_1(z)/dz + a_2(z)df_2(z)/dz. \qquad (2.3.4b)$$

We note that the derivative of ψ is obtained *formally* as if $a_1(z)$ and $a_2(z)$ were constants, which one achieves by imposing on $a_1(z)$ and $a_2(z)$ the condition

$$f_1(z)da_1(z)/dz + f_2(z)da_2(z)/dz = 0. \qquad (2.3.4c)$$

Using (2.3.1a,b), (2.3.2), (2.3.3) and (2.3.4a–c), we can replace the differential equation (2.1.1) by a system of two coupled differential equations of the first order for $a_1(z)$ and $a_2(z)$, which we write in matrix form as

$$\frac{d}{dz}\begin{pmatrix} a_1(z) \\ a_2(z) \end{pmatrix} = \mathbf{M}(z)\begin{pmatrix} a_1(z) \\ a_2(z) \end{pmatrix}, \qquad (2.3.5a)$$

$$\mathbf{M}(z) = \frac{1}{2}i\varepsilon(z)q(z)\begin{pmatrix} 1 & \exp[-2i\,w(z)] \\ -\exp[+2i\,w(z)] & -1 \end{pmatrix}, \qquad (2.3.5b)$$

$$\varepsilon = q^{-3/2}d^2q^{-1/2}/dz^2 + (R - q^2)/q^2. \qquad (2.3.5c)$$

If ε is identically equal to zero, (2.3.5c) agrees with the q-equation (2.1.4), and it follows from (2.3.5a,b) that a_1 and a_2 are constant, and hence $f_1(z)$ and $f_2(z)$ are exact solutions of the differential equation (2.1.1). If ε is small in a certain local region of the complex z-plane, a_1 and a_2 are approximately constant in that region, and the functions $f_1(z)$ and $f_2(z)$ satisfy the differential equation approximately, which means that $f_1(z)$ and $f_2(z)$ solve the *local* problem in that region. In order that $a_1(z)$ and $a_2(z)$ be locally approximately constant, it is thus necessary that the quantity ε is small. This is, however, not a sufficient condition for $a_1(z)$ and $a_2(z)$ to be approximately constant in a *global* region

The differential equation (2.3.5a) can be replaced by the integral equation

$$
\begin{pmatrix} a_1(z) \\ a_2(z) \end{pmatrix} = \begin{pmatrix} a_1(z_0) \\ a_2(z_0) \end{pmatrix} + \int_{z_0}^{z} dz \mathbf{M}(z) \begin{pmatrix} a_1(z) \\ a_2(z) \end{pmatrix}, \qquad (2.3.6)
$$

the solution of which can be obtained in closed form by an iteration procedure yielding

$$
\begin{pmatrix} a_1(z) \\ a_2(z) \end{pmatrix} = \mathbf{F}(z, z_0) \begin{pmatrix} a_1(z_0) \\ a_2(z_0) \end{pmatrix}, \qquad (2.3.7)
$$

where $\mathbf{F}(z, z_0)$ is a two-by-two matrix given by the convergent series

$$
\mathbf{F}(z, z_0) = \mathbf{I} + \int_{z_0}^{z} dz_1 \mathbf{M}(z_1) + \int_{z_0}^{z} dz_1 \mathbf{M}(z_1) \int_{z_0}^{z_1} dz_2 \mathbf{M}(z_2)
$$

$$
+ \int_{z_0}^{z} dz_1 \mathbf{M}(z_1) \int_{z_0}^{z_1} dz_2 \mathbf{M}(z_2) \int_{z_0}^{z_2} dz_3 \mathbf{M}(z_3) + \cdots, \qquad (2.3.8)
$$

$\mathbf{I}$ being the two-by-two unit matrix. The formula (2.3.8) would be of limited value for our purpose if it were not so that the elements of the product matrix $\mathbf{M}(z_1)\,\mathbf{M}(z_2)\cdots\mathbf{M}(z_n)$, appearing in the general term on the right-hand side of (2.3.8), can be factorized as follows

$$
\mathbf{M}(z_1)\,\mathbf{M}(z_2)\cdots\mathbf{M}(z_n) = \left(\frac{1}{2}i\right)^n \varepsilon(z_1)q(z_1)\varepsilon(z_2)q(z_2)\cdots\varepsilon(z_n)q(z_n)
$$

$$
\times (1 - \exp\{-2i[w(z_1) - w(z_2)]\})
$$
$$
\times (1 - \exp\{-2i[w(z_2) - w(z_3)]\})\cdots
$$
$$
\times (1 - \exp\{-2i[w(z_{n-1}) - w(z_n)]\})
$$
$$
\times \begin{pmatrix} 1 & \exp[-2iw(z_n)] \\ -\exp[2iw(z_1)] & -\exp\{2i[w(z_1) - w(z_n)]\} \end{pmatrix}.
$$

$$(2.3.9)$$

This is an important fact which makes it possible for us to obtain explicit expressions for the elements of the matrix $\mathbf{F}(z, z_0)$ as fairly simple, convergent series. Substituting (2.3.9) into (2.3.8), we thus obtain (Fröman and Fröman 1965)

$$
F_{11}(z, z_0) = 1 + \int_{z_0}^{z} dz_1 \tfrac{1}{2}i\varepsilon(z_1)q(z_1) + \cdots, \qquad (2.3.10a)
$$

$$F_{12}(z, z_0) = \int_{z_0}^{z} dz_1 \tfrac{1}{2} i\varepsilon(z_1) q(z_1) \exp[-2i\, w(z_1)] + \cdots, \qquad (2.3.10\text{b})$$

$$F_{21}(z, z_0) = -\int_{z_0}^{z} dz_1 \tfrac{1}{2} i\varepsilon(z_1) q(z_1) \exp[2i\, w(z_1)] + \cdots, \qquad (2.3.10\text{c})$$

$$F_{22}(z, z_0) = 1 - \int_{z_0}^{z} dz_1 \tfrac{1}{2} i\varepsilon(z_1) q(z_1) + \cdots. \qquad (2.3.10\text{d})$$

2.3.2 General relations satisfied by the F-matrix

The matrix $\mathbf{F}(z, z_0)$ satisfies the differential equation

$$\frac{\partial}{\partial z}\mathbf{F}(z, z_0) = \mathbf{M}(z)\mathbf{F}(z, z_0) \qquad (2.3.11)$$

and is in fact the particular solution of this differential equation that is equal to the two-by-two unit matrix for $z = z_0$. The F-matrix also satisfies the following general relations

$$\det \mathbf{F}(z, z_0) = 1, \qquad (2.3.12\text{a})$$
$$\mathbf{F}(z, z_0) = \mathbf{F}(z, z_1)\mathbf{F}(z_1, z_0), \qquad (2.3.12\text{b})$$
$$\mathbf{F}(z_0, z) = [\mathbf{F}(z, z_0)]^{-1} = \begin{pmatrix} F_{22}(z, z_0) & -F_{12}(z, z_0) \\ -F_{21}(z, z_0) & F_{11}(z, z_0) \end{pmatrix}. \qquad (2.3.12\text{c})$$

It follows from (2.3.11) and (2.3.5b) that

$$\exp[-2iw(z)]\frac{\partial F_{21}(z, z_0)}{\partial z} + \frac{\partial F_{11}(z, z_0)}{\partial z} = 0, \qquad (2.3.13\text{a})$$

$$\exp[-2iw(z)]\frac{\partial F_{22}(z, z_0)}{\partial z} + \frac{\partial F_{12}(z, z_0)}{\partial z} = 0, \qquad (2.3.13\text{b})$$

and from (2.3.13a,b) and (2.3.1c) that

$$\frac{\partial}{\partial z}\{\exp[-2iw(z)]F_{21}(z, z_0) + F_{11}(z, z_0)\} = -2iq(z)\exp[-2iw(z)]F_{21}(z, z_0),$$
$$(2.3.14\text{a})$$

$$\frac{\partial}{\partial z}\{\exp[-2iw(z)]F_{22}(z, z_0) + F_{12}(z, z_0)\} = -2iq(z)\exp[-2iw(z)]F_{22}(z, z_0).$$
$$(2.3.14\text{b})$$

The general properties of the functions $R(z)$ and $Q^2(z)$ are reflected in the properties of the F-matrix. Thus, symmetry relations between elements of the F-matrix exist in the special case when $R(z)$ and $Q^2(z)$ are real on the real z-axis and the points z and z_0 both lie on this axis. If $R(-z) = R(z)$ and $Q^2(-z) = Q^2(z)$, or if $R(z)$ and $Q^2(z)$ are periodic functions of z with the same period, other symmetry relations exist; see Fröman, Fröman and Lundborg (1988b).

2.3.3 F-matrix corresponding to the encircling of a simple zero of $Q^2(z)$

We shall now give an *exact* formula for the F-matrix connecting any point z with the corresponding point z' lying on the next Riemann sheet and obtained by moving from z in the *positive* sense along a path encircling a simple zero t of $Q^2(z)$. The function $Q^2(z)$ need not be real on the real axis, and the zero t may lie anywhere in the complex z-plane. Defining $w(z)$ according to (2.2.15), we have $q^{-1/2}(z') = -iq^{-1/2}(z)$ and $w(z') = -w(z)$, and hence $f_1(z') = -if_2(z)$ and $f_2(z') = -if_1(z)$. Recalling (2.3.4a,b), we therefore get

$$\psi(z') = a_1(z')f_1(z') + a_2(z')f_2(z') = -ia_2(z')f_1(z) - ia_1(z')f_2(z), \quad (2.3.15a)$$
$$d\psi(z')/dz' = a_1(z')f_1'(z') + a_2(z')f_2'(z') = -ia_2(z')f_1'(z) - ia_1(z')f_2'(z). \quad (2.3.15b)$$

The solution ψ is single-valued in the region around the transition zero t, provided that there is no pole of $R(z)$ in this region. Comparing (2.3.15a,b) with (2.3.4a,b), we obtain $a_1(z) = -ia_2(z')$ and $a_2(z) = -ia_1(z')$, which according to (2.3.7) implies that

$$\mathbf{F}(z', z) = \begin{pmatrix} 0 & i \\ i & 0 \end{pmatrix}. \quad (2.3.16)$$

2.3.4 Basic estimates

Useful estimates of the elements of $\mathbf{F}(z, z_0)$ have been derived from the series in (2.3.10a–d) by Fröman and Fröman (1965) on the assumption that the points z and z_0 can be connected by a path in the complex z-plane along which the absolute value of $\exp[i\,w(z)]$ increases monotonically (in the non-strict sense) in the direction from z_0 to z. These estimates, which we call *basic estimates*, can be written as

$$|F_{11}(z, z_0) - 1| \le \frac{1}{2}\mu + \text{higher powers of } \mu, \quad (2.3.17a)$$

$$|F_{12}(z, z_0)| \le |\exp\{-2iw(z_0)\}|\left(\frac{1}{2}\mu + \text{higher powers of } \mu\right), \quad (2.3.17b)$$

$$|F_{21}(z, z_0)| \leq |\exp\{2iw(z)\}| \left(\frac{1}{2}\mu + \text{higher powers of } \mu\right), \quad (2.3.17c)$$

$$|F_{22}(z, z_0) - 1| \leq \frac{1}{2}\mu + |\exp\{2i[w(z) - w(z_0)]\}| \left(\frac{1}{4}\mu^2 + \text{ higher powers of } \mu\right),$$

$$(2.3.17d)$$

where

$$\mu = \mu(z, z_0) = \left| \int_{z_0}^{z} |\varepsilon(z)q(z)dz| \right| \quad (2.3.18)$$

with the integration performed along the path of monotonicity. When we are using the estimates (2.3.17a–d), we shall always assume that $\mu \ll 1$, which is in general the case if $R(z)$ varies slowly, if $Q(z)$ is chosen conveniently and if the path of monotonicity does not pass close to a transition point where the phase-integral approximation breaks down. The μ-integral (2.3.18) plays an important role as a measure of the 'goodness' of the phase-integral approximation. If the points Z, z_0 and z lie in the same *classically allowed region* of the real z-axis, $w(z)$ and $w(z_0)$ are real, and when $\mu \ll 1$, the matrix $\mathbf{F}(z, z_0)$ is approximately a unit matrix, and both $a_1(z)$ and $a_2(z)$ are approximately constant. On the other hand, in a *classically forbidden region* of the real z-axis the absolute value of $\exp[iw(z)]$ increases or decreases rapidly, and some of the elements of the F-matrix become large. Therefore, one of the coefficients $a_1(z)$ and $a_2(z)$ may change violently. The rather delicate properties of the phase-integral approximation along a path where the absolute value of $\exp[iw(z)]$ increases or decreases rapidly have been analysed in detail by N. Fröman (1966a).

When $\mu \ll 1$ and

$$|\exp[iw(z_0)]| \approx |\exp[iw(z)]| \approx 1, \quad (2.3.19)$$

the basic estimates (2.3.17a–d) provide approximate values of all four elements of the matrix $\mathbf{F}(z, z_0)$, which in this case is seen to be approximately equal to the two-by-two unit matrix. In other cases, the basic estimates do not provide approximate values of all elements of $\mathbf{F}(z, z_0)$, but one can still use the estimates (2.3.17a–d) for handling connection problems.

In physical applications, the general procedure when one uses the F-matrix method is to derive a convenient exact expression for the physical quantity under consideration and then to omit small quantities expressed in terms of elements of the F-matrix. In general situations when the points z_0 and z cannot be joined by a path along which the absolute value of $\exp[iw(z)]$ is monotonic, one has to divide the path into monotonic parts, use the relations (2.3.12a–c) and possibly other existing relations to bring the expression for the physical quantity of interest into a convenient form, and then use the basic estimates (2.3.17a–d) to obtain upper bounds for the terms to be neglected. For the simplest non-monotonic path this can be done as follows.

Consider two points z' and z'' and a path joining these points in the complex z-plane. On this path the absolute value of the function $\exp[iw(z)]$ is assumed to have a single maximum or minimum at a point z_m, and the absolute value of $\exp[-iw(z)]$, in the case of a maximum, or $\exp[+iw(z)]$, in the case of a minimum, is assumed to be not too large at the points z' and z''. It is also assumed that in the case of a maximum $|F_{21}(z', z'')|$ is not too large, and in the case of a minimum $|F_{12}(z', z'')|$ is not too large. Furthermore, the μ-integral (2.3.18) for the whole path from z' to z'' is assumed to be small compared to unity. Using (2.3.12a,b) and (2.3.17a–d), we then obtain the following approximate formulas corresponding to a minimum and to a maximum, respectively, at z_m (Fröman, Fröman and Lundborg 1988a):

$$\mathbf{F}(z', z'') = \begin{pmatrix} 1 & S' \\ 0 & 1 \end{pmatrix}, \tag{2.3.20a}$$

$$\mathbf{F}(z', z'') = \begin{pmatrix} 1 & 0 \\ S'' & 1 \end{pmatrix}, \tag{2.3.20b}$$

where S' and S'' are so far unknown Stokes constants. Formulas (2.3.20a,b) have important applications.

2.3.5 Stokes and anti-Stokes lines

We have seen above that, when tracing a solution in the complex z-plane, one must know how the monotonicity of the absolute value of $\exp[iw(z)]$ changes along the path. This can be found from the pattern of the Stokes and anti-Stokes lines, which are defined by the conditions

$$dw = q(z)dz \text{ is purely imaginary along a Stokes line,} \tag{2.3.21a}$$

$$dw = q(z)dz \text{ is real along an anti-Stokes line.} \tag{2.3.21b}$$

Thus, the Stokes lines are lines along which the absolute value of $\exp[iw(z)]$ increases or decreases most rapidly, while the anti-Stokes lines are level lines for constant absolute values of $\exp[iw(z)]$. At points where $q(z)$ is regular and different from zero the Stokes and anti-Stokes lines are orthogonal to each other; see Section 3.4. The importance of the Stokes and anti-Stokes lines for tracing a solution of the differential equation in the complex z-plane will be further clarified in Section 3.5 and 3.6. If one knows the pattern of the Stokes and anti-Stokes lines in the neighbourhood of the transition points, it is in general not difficult to get their overall pattern.

When $w(z)$ is given by (2.2.15), where t is a transition point of odd order, we *define* a Stokes line *emerging* from t as a Stokes line on which $w(z)$ is purely imaginary, and we *define* an anti-Stokes line *emerging* from t as an anti-Stokes line on which $w(z)$ is real. For the first-order approximation it is easily seen that these lines start at the transition point t itself.

2.3.6 Symbols facilitating the tracing of a wave function in the complex z-plane

When the pattern of the Stokes and anti-Stokes lines in the complex z-plane is known, the tracing of a wave function, represented in different parts of the plane by different linear combinations of the phase-integral functions, can be facilitated by means of convenient symbols displaying the directions in which the absolute value and the phase of the exponentials $\exp[+iw(z)]$ and $\exp[-iw(z)]$ increase along the Stokes and anti-Stokes lines.

Consider the Stokes line and the anti-Stokes line through a given point, where $q(z)$ is regular and different from zero. At this point we associate with each one of the exponentials $\exp[+iw(z)]$ and $\exp[-iw(z)]$, and hence with each one of the phase-integral functions $f_1(z)$ and $f_2(z)$ defined by (2.3.1a–c), a compound symbol consisting of a simple arrow along the Stokes line in the direction of increasing absolute value of the exponential to which it belongs, and a double arrow along the anti-Stokes line in the direction of increasing phase for the same exponential; see Fig. 2.3.1(a). The direction of the double arrow is always obtained by a rotation of $\pi/2$ in the positive direction relative to the direction of the simple arrow. The linear combination $a_1 f_1(z) + a_2 f_2(z)$, with both terms assumed to be significant,

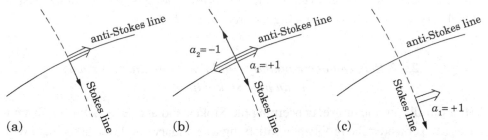

Figure 2.3.1 (a) The compound symbol that can be associated with either of the phase-integral functions $f_1(z)$ and $f_2(z)$. (b) The two compound symbols that represent the linear combination $f_1(z) - f_2(z)$. (c) The compound symbol that represents the wave function in (b) when z has moved along the Stokes line such that the term $-f_2(z)$ has become insignificant.

can then be represented graphically in the complex z-plane, at any point where $q(z)$ is regular and different from zero, by the two compound symbols in Fig. 2.3.1(b) with appropriate orientation and with the actual values of a_1 and a_2 indicated, as shown in Fig. 2.3.1(b) for the case when $a_1 = +1$ and $a_2 = -1$. When only one of the two terms in the linear combination is significant, we delete the compound symbol corresponding to the insignificant term; see Fig. 2.3.1(c).

The compound symbols in Fig. 2.3.1 constitute 'vehicles' for the movement of which in the complex z-plane one can formulate certain 'traffic rules'. These rules are based on the behaviour of the phase-integral solution along Stokes lines (Section 3.6) and anti-Stokes lines (Section 3.5), and on how the a-coefficients change when z moves from one anti-Stokes line to a neighbouring one, both emerging from a simple zero of $Q^2(z)$ (see Section 3.7), or more generally from a first-order pole or an odd-order zero of $Q^2(z)$; see sections 7.3 and 7.4 in Fröman and Fröman (1965). Along an anti-Stokes line the vehicle can move in both directions, and the coefficients a_1 and a_2 both remain approximately constant; see Section 3.5. When the vehicle moves along a Stokes line, the a-coefficient associated with the simple arrow in the direction of motion remains approximately constant, while the term associated with the other a-coefficient rapidly becomes insignificant and should then be deleted; see Section 3.6. Thus, when for instance the term with the coefficient a_2 becomes insignificant, the symbol in Fig. 2.3.1(b) is replaced by the symbol in Fig. 2.3.1(c). Using these 'traffic rules' one can easily trace a desired solution of the Schrödinger equation along Stokes and anti-Stokes lines in the complex z-plane. One can also easily find the orientation on neighbouring Stokes and anti-Stokes lines, emerging from the same simple pole or odd-order zero of $Q^2(z)$, of the compound symbol corresponding to a given phase-integral function; see Fig. 2.3.2. To find how the a-coefficients change when z moves from an anti-Stokes line to a neighbouring one, both of which emerge from a simple zero of $Q^2(z)$, one uses (3.7.3a–c) and (3.7.6) or (3.7.7a–c) and (3.7.8) in Section 3.7. In sections 7.3 and 7.4 of Fröman and Fröman (1965) there are corresponding formulas for the case in which t is a first-order pole or a higher-order zero of $Q^2(z)$.

2.3.7 Removal of a boundary condition from the real z-axis to an anti-Stokes line

After our discussion above, concerning the Stokes and anti-Stokes lines, it is convenient to consider the problem of satisfying a boundary condition associated with quasi-stationary states. To calculate the energy levels of such states, for instance in a radial problem, one allows the energy E to have a small, negative imaginary part and requires that the wave function is represented by a purely outgoing wave as $z \to +\infty$. The absolute value of the quasi-stationary wave function is then

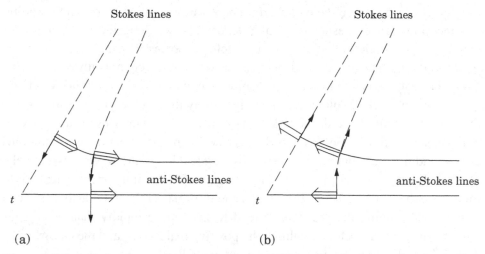

Figure 2.3.2 A Stokes line and a neighbouring anti-Stokes line, both emerging from a first-order zero of $Q^2(z)$. According to the 'traffic rules' the compound symbol depicted in (a) can move from the Stokes line to the anti-Stokes line, and the compound symbol in (b) can move from the anti-Stokes line to the Stokes line. In this way one finds the orientation of each compound symbol on two neighbouring Stokes and anti-Stokes lines.

exponentially increasing as z tends to $+\infty$, and to satisfy the boundary condition *on the real axis* at $z \to +\infty$ one cannot use the phase-integral approximation, since on this axis (when z is sufficiently large) the subdominant term is small compared with the error of the dominant term (because of the small, negative imaginary part of E). One can handle this difficulty by moving the boundary condition from the real z-axis to the anti-Stokes line that emerges from the transition zero t lying farthest to the right and on which Re z increases towards $+\infty$ as z moves away from t. A concrete application of this procedure by Fröman, Fröman, Andersson and Hökback (1992) concerns the treatment of a problem involving black-hole normal modes. Another application appears in Section 3.48.

To explain in more detail what has been said above, we consider two *exact* solutions of the forms:

$$\psi_+ = \chi_+(z)\exp(+ikz), \tag{2.3.22a}$$
$$\psi_- = \chi_-(z)\exp(-ikz), \tag{2.3.22b}$$

where

$$k = (2mE/\hbar^2)^{1/2} \tag{2.3.23}$$

and

$$\chi_\pm(z) \to k^{-1/2}, \tag{2.3.24a}$$
$$d\chi_\pm(z)/dz \to 0, \tag{2.3.24b}$$

as z tends to $+\infty$ *along the real z-axis*. These *exact* solutions represent outgoing and incoming waves, respectively, as z tends to $+\infty$ along the real z-axis. We now assume that the potential $V(z)$ has such properties that the *exact* functions $\chi_+(z)$ and $\chi_-(z)$ tend to $k^{-1/2}$ and their derivatives tend to zero not only as $z \to +\infty$ along the real z-axis, but also as the absolute value of z tends to infinity when arg z lies in an interval of finite size. The point z may then lie on the anti-Stokes line that emerges from t towards the right and on which Re z increases towards $+\infty$ as z moves away from t. On this anti-Stokes line the phase-integral approximations corresponding to ψ_+ and ψ_- represent outgoing and incoming waves, respectively. One can thus move the original boundary condition from infinity on the real z-axis to infinity on the anti-Stokes line in question and thus master the previously described difficulty. Assuming the potential to tend to zero more rapidly than const$/z$ as $z \to \infty$ in a certain sector containing the positive real z-axis, and the energy to be $E = E^{(0)} - i\Gamma/2$ (see Section 3.48), we obtain as Re $z \to \infty$

$$
\int^z q(z)dz \approx \int^z (2mE/\hbar^2)^{1/2}dz = \text{const} + \left[2m\left(E^{(0)} - \tfrac{1}{2}i\Gamma\right)\big/\hbar^2\right]^{1/2}z
$$

$$
\approx \text{const} + \left(\frac{2mE^{(0)}}{\hbar^2}\right)^{1/2}\left(1 - \frac{i\Gamma}{4E^{(0)}}\right)(x + iy)
$$

$$
= \text{const} + \left(\frac{2mE^{(0)}}{\hbar^2}\right)^{1/2}\left[\left(x + \frac{\Gamma}{4E^{(0)}}y\right) + i\left(y - \frac{\Gamma}{4E^{(0)}}x\right)\right],
$$

and the anti-Stokes lines far to the right of t are thus given by

$$
y = x\Gamma\big/\left(4E^{(0)}\right) + \text{const}, \tag{2.3.25}
$$

and, for sufficiently large values of $x(= \text{Re}z)$, they lie in the upper half of the complex z-plane. Since Γ is very small compared with $E^{(0)}$, it is seen from (2.3.25) that the anti-Stokes lines in question are finally (as Re $z \to +\infty$) almost parallel to the real axis.

2.3.8 Dependence of the F-matrix on the lower limit of integration in the phase integral

In the definition (2.3.1c) of $w(z)$ the point Z appears as the lower limit of integration. If we change the position of this point, the value of $w(z)$ will be changed by an additive constant, which will affect the phase-integral functions $f_1(z)$ and $f_2(z)$ and the non-diagonal elements of the matrix $\mathbf{F}(z, z_0)$. Since it is convenient to have explicit expressions for this dependence on Z of the F-matrix, we shall now relate quantities pertaining to the choice Z in (2.3.1c) to quantities pertaining to the choice $\bar{Z}$, each one of the latter quantities being marked by a bar. Thus we introduce the phase integral

$$\bar{w}(z) = \int\limits_{(\bar{Z})}^{z} q(z)dz,$$ (2.3.26)

which is related to the phase integral $w(z)$ in (2.3.1c) as follows:

$$w(z) = \bar{w}(z) + \omega,$$ (2.3.27a)

$$\omega = \int\limits_{(Z)}^{\bar{Z}} q(z)dz.$$ (2.3.27b)

The relation between $\mathbf{F}(z, z_0)$ and $\bar{\mathbf{F}}(z, z_0)$ is easily obtained if we note that, according to the convergent series expansions (2.3.10a–d), the quantities $F_{11}(z, z_0)$, $F_{12}(z, z_0)\exp(2iw)$, $F_{21}(z, z_0)\exp(-2iw)$ and $F_{22}(z, z_0)$, where w stands for either $w(z)$ or $w(z_0)$, are all independent of the choice of Z in the integral (2.3.1c) defining $w(z)$. We thus obtain, with due regard to (2.3.27a), the transformation formula

$$\mathbf{F}(z, z_0) = \begin{pmatrix} \bar{F}_{11}(z, z_0) & \bar{F}_{12}(z, z_0)\exp(-2i\omega) \\ \bar{F}_{21}(z, z_0)\exp(+2i\omega) & \bar{F}_{22}(z, z_0) \end{pmatrix}$$ (2.3.28)

$$= \begin{pmatrix} \exp(-i\omega) & 0 \\ 0 & \exp(+i\omega) \end{pmatrix} \bar{\mathbf{F}}(z, z_0) \begin{pmatrix} \exp(+i\omega) & 0 \\ 0 & \exp(-i\omega) \end{pmatrix}.$$
(2.3.28')

2.3.9 F-matrix expressed in terms of two linearly independent solutions of the differential equation

We shall now show how one can obtain the F-matrix from two linearly independent solutions $\psi(z)$ and $\bar{\psi}(z)$ of the differential equation (2.1.1). Denoting the corresponding a-coefficients by $a_1(z)$, $a_2(z)$ and $\bar{a}_1(z)$, $\bar{a}_2(z)$, respectively, we have according to (2.3.7) the formulas

$$\begin{pmatrix} a_1(z) \\ a_2(z) \end{pmatrix} = \mathbf{F}(z, z_0) \begin{pmatrix} a_1(z_0) \\ a_2(z_0) \end{pmatrix},$$ (2.3.29a)

$$\begin{pmatrix} \bar{a}_1(z) \\ \bar{a}_2(z) \end{pmatrix} = \mathbf{F}(z, z_0) \begin{pmatrix} \bar{a}_1(z_0) \\ \bar{a}_2(z_0) \end{pmatrix},$$ (2.3.29b)

which can be summarized into the formula

$$\begin{pmatrix} a_1(z) & \bar{a}_1(z) \\ a_2(z) & \bar{a}_2(z) \end{pmatrix} = \mathbf{F}(z, z_0) \begin{pmatrix} a_1(z_0) & \bar{a}_1(z_0) \\ a_2(z_0) & \bar{a}_2(z_0) \end{pmatrix},$$ (2.3.30)

from which we obtain

$$\mathbf{F}(z, z_0) = \begin{pmatrix} a_1(z) & \bar{a}_1(z) \\ a_2(z) & \bar{a}_2(z) \end{pmatrix} \begin{pmatrix} a_1(z_0) & \bar{a}_1(z_0) \\ a_2(z_0) & \bar{a}_2(z_0) \end{pmatrix}^{-1}. \tag{2.3.31}$$

From (2.3.4a,b) and the corresponding formulas for the quantities with bars we get

$$\begin{pmatrix} \psi(z) & \bar{\psi}(z) \\ \psi'(z) & \bar{\psi}'(z) \end{pmatrix} = \begin{pmatrix} f_1(z) & f_2(z) \\ f_1'(z) & f_2'(z) \end{pmatrix} \begin{pmatrix} a_1(z) & \bar{a}_1(z) \\ a_2(z) & \bar{a}_2(z) \end{pmatrix}, \tag{2.3.32}$$

and from this formula, with z replaced by z_0, we obtain with the aid of (2.3.2)

$$\begin{vmatrix} a_1(z_0) & \bar{a}_1(z_0) \\ a_2(z_0) & \bar{a}_2(z_0) \end{vmatrix} = -\frac{1}{2i} \begin{vmatrix} \psi(z_0) & \bar{\psi}(z_0) \\ \psi'(z_0) & \bar{\psi}'(z_0) \end{vmatrix}$$

$$= -\frac{\psi(z_0)\bar{\psi}'(z_0) - \bar{\psi}(z_0)\psi'(z_0)}{2i}. \tag{2.3.33}$$

Using (2.3.33), we obtain

$$\begin{pmatrix} a_1(z_0) & \bar{a}_1(z_0) \\ a_2(z_0) & \bar{a}_2(z_0) \end{pmatrix}^{-1} = \begin{vmatrix} a_1(z_0) & \bar{a}_1(z_0) \\ a_2(z_0) & \bar{a}_2(z_0) \end{vmatrix}^{-1} \begin{pmatrix} \bar{a}_2(z_0) & -\bar{a}_1(z_0) \\ -a_2(z_0) & a_1(z_0) \end{pmatrix}$$

$$= -\frac{2i}{\psi(z_0)\bar{\psi}'(z_0) - \bar{\psi}(z_0)\psi'(z_0)} \begin{pmatrix} \bar{a}_2(z_0) & -\bar{a}_1(z_0) \\ -a_2(z_0) & a_1(z_0) \end{pmatrix}, \tag{2.3.34}$$

and therefore (2.3.31) gives

$$\mathbf{F}(z, z_0) = -\frac{2i}{\psi(z_0)\bar{\psi}'(z_0) - \bar{\psi}(z_0)\psi'(z_0)}$$

$$\times \begin{pmatrix} a_1(z)\bar{a}_2(z_0) - \bar{a}_1(z)a_2(z_0) & -a_1(z)\bar{a}_1(z_0) + \bar{a}_1(z)a_1(z_0) \\ a_2(z)\bar{a}_2(z_0) - \bar{a}_2(z)a_2(z_0) & -a_2(z)\bar{a}_1(z_0) + \bar{a}_2(z)a_1(z_0) \end{pmatrix}. \tag{2.3.35}$$

To obtain the a-coefficients in (2.3.35) we solve (2.3.4a,b) for $a_1(z)$ and $a_2(z)$, getting with the aid of (2.3.2)

$$a_1(z) = -\frac{1}{2}i[f_2(z)\psi'(z) - f_2'(z)\psi(z)], \tag{2.3.36a}$$

$$a_2(z) = +\frac{1}{2}i[f_1(z)\psi'(z) - f_1'(z)\psi(z)]. \tag{2.3.36b}$$

Similarly we obtain

$$\bar{a}_1(z) = -\frac{1}{2}i[f_2(z)\bar{\psi}'(z) - f_2'(z)\bar{\psi}(z)], \tag{2.3.37a}$$

$$\bar{a}_2(z) = +\frac{1}{2}i[f_1(z)\bar{\psi}'(z) - f_1'(z)\bar{\psi}(z)]. \tag{2.3.37b}$$

The formulas in this section form the basis for the parameterization in Section 2.4 of the F-matrix connecting points on opposite sides of a turning point (on the real axis) and are also essential for the solution of several of the problems in Chapter 3.

2.4 F-matrix connecting points on opposite sides of a well-isolated turning point, and expressions for the wave function in these regions

We recall that we are using the term turning point in a generalized sense to designate a simple zero of $Q^2(z)$ on the real z-axis, when $R(z)$ and $Q^2(z)$ are real on this axis. We also use the terms classically allowed and classically forbidden region in a correspondingly generalized sense to designate regions on the real axis (the x-axis) where $Q^2(x) > 0$ and $Q^2(x) < 0$, respectively. A turning point is said to be well isolated when it lies far away from other transition points.

In this section we shall be concerned with the connection of phase-integral solutions across a turning point. For this case the F-matrix in the rigorous method for solving connection problems, briefly presented in Section 2.3, can be parameterized such that the exact wave function on each side of a turning point is obtained in a form that displays its properties in a way that is very useful in applications where subtle effects are to be accounted for. The parameterization of the F-matrix connecting points on opposite sides of a turning point, which is presented here for the first time, was worked out mainly for the calculation of the exponentially small change of the phase of the wave function in a classically allowed region, due to a change of the boundary condition in an adjacent classically forbidden region. One cannot make this calculation in a satisfactory way by means of the connection formulas associated with a turning point because of their one-directional nature. The smallness of the change of the phase shift in question requires a sophisticated treatment with complete control of the errors in the course of the calculation. For this purpose the F-matrix method, as used so far, has proved to be efficient, but to involve rather cumbersome and non-trivial manipulations; see Fröman, Yngve and Fröman (1987). The parameterization of the F-matrix facilitates the treatment greatly, although the essential ideas are the same as in Fröman and Fröman (1965) and in Fröman, Yngve and Fröman (1987). The intricate problem treated in the last-mentioned paper, i.e., the calculation of the exponentially small change of the energy levels when the radial wave function of an atom is required to be equal to zero at a finite value of r instead of at infinity (compressed atom), can be treated in a simple and straightforward way by means of the parameterized F-matrix; see Section 3.29.

We consider the matrix $\mathbf{F}(x_1, x_2)$, where x_1 and x_2 are points on the real axis in the classically forbidden and classically allowed regions, respectively, on opposite sides of a well-isolated turning point. For the common situation that $R(z)$ as well as $Q^2(z)$ is real on the real z-axis, we introduce three judiciously chosen real parameters $\alpha(x_1, x_2)$, $\beta(x_1, x_2)$ and $\gamma(x_1, x_2)$ in terms of which the elements of $\mathbf{F}(x_1, x_2)$ are expressed with their real and imaginary parts explicitly displayed. We obtain useful estimates for α, β and γ, when x_1 and x_2 lie sufficiently far away from the turning point, and also for the changes of these parameters, when x_1 and x_2 are displaced along the real axis in the classically forbidden and classically allowed regions, respectively. These estimates are crucial for the analysis of delicate problems, such as a compressed atom. In this section we also give expressions for the *exact wave function* on each side of the turning point in terms of the phase-integral quantities and the parameters α, β and γ. The estimates of α, β and γ easily yield the connection formulas and at the same time display their one-directional nature, as will be shown in some problems. With the aid of the parameterized F-matrix we shall in Section 3.30 explain the strange circumstance that an essentially correct result can *sometimes* be obtained by the use of connection formulas in forbidden directions, and there we shall also clarify when quite erroneous results are obtained in that way; see also Subsection 2.4.4.

By means of the parameterized matrix $\mathbf{F}(x_1, x_2)$ the *exact* wave function at the points x_1 (in the classically forbidden region) and x_2 (in the classically allowed region) is expressed in such a form that the cumbersome work with the phases of $q(z)$ and $w(z)$ and with different cases depending on the choice of these phases is simplified considerably. In practical applications one can thus work with the *exact* wave function, and at the end of the calculations one can introduce in a quite controllable way the approximations of the parameters α, β and γ. Thus one obtains a reliable result in a straightforward way.

2.4.1 Symmetry relations and estimates of the F-matrix elements

We assume that the coefficient function $R(z)$ in the differential equation (2.1.1) is real on the real z-axis, and we also choose $Q^2(z)$ to be real there. On the part of the real z-axis under consideration there is a turning point t, i.e., a simple zero of $Q^2(z)$. We may have to the left of t a classically forbidden region and to the right of t a classically allowed region, as shown in Fig. 2.4.1, or vice versa, as shown in Fig. 2.4.2. An unspecified point on the real z-axis (lying not too close to t) is denoted by x_1 in the classically forbidden region and by x_2 in the classically allowed region. Thus we have $Q^2(x_1) < 0$ and $Q^2(x_2) > 0$. Since $R(z)$ and $Q^2(z)$ are real on the real z-axis, there exist certain relations, called symmetry relations, between the elements of the matrix $\mathbf{F}(x_1, x_2)$. These symmetry relations are a consequence of the fact that if $\psi(z)$ is a solution of the differential equation (2.1.1) on the real z-axis, so also is the complex conjugate wave function $\psi^*(z)$, because of the assumption that $R(z)$ is real on the the real z-axis. For a detailed derivation of this we refer the reader to Fröman and Fröman (1965) and Fröman, Fröman and Lundborg (1988b).

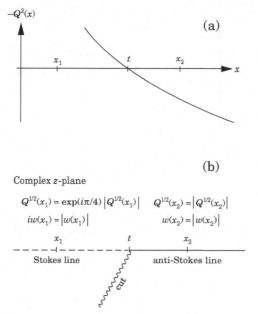

Figure 2.4.1 When there is a classically allowed region to the right of the turning point t, (a) gives a schematic picture of the function $-Q^2(x)$, and (b) shows the phase of $Q^{1/2}(x)$.

When we choose $Q^{1/2}(x_2) = |Q^{1/2}(x_2)|$, and $w(z)$ is defined as

$$w(z) = \int\limits_{(t)}^{z} q(z)dz, \qquad (2.4.1)$$

where we have used the short-hand notation explained in (2.2.14) and (2.2.15), the symmetry relations in question are (Fröman and Fröman 1965, eqs. (6.10a,b))

$$F_{12}(x_1, x_2) = \pm i F_{11}^*(x_1, x_2), \qquad (2.4.2a)$$

$$F_{21}(x_1, x_2) = \pm i F_{22}^*(x_1, x_2), \qquad (2.4.2b)$$

where the upper and lower signs refer to the cases in Fig. 2.4.1 and Fig. 2.4.2, respectively. Note that the cut in Fig. 2.4.1 as well as that in Fig. 2.4.2 is introduced in such a way that we move from x_1 to x_2, or vice versa, in the *upper* half of the complex z-plane. One can then find a path in the complex z-plane, which joins the points x_1 and x_2, on which the absolute value of $\exp[i w(z)]$ has a single minimum. With the aid of the multiplication rule (2.3.12b), the symmetry relations (2.4.2a,b) and the basic estimates (2.3.17a,c) one then obtains the estimates (Fröman and Fröman 1965, eqs. (6.13a,b))

$$|F_{11}(x_1, x_2) - 1| \leq \mu + \text{higher powers of } \mu, \qquad (2.4.3a)$$

$$|F_{22}(x_1, x_2)| \leq |\exp[2iw(x_1)]| \left(\tfrac{1}{2}\mu + \text{higher powers of } \mu\right), \qquad (2.4.3b)$$

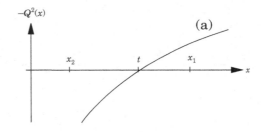

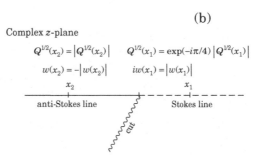

Figure 2.4.2 When there is a classically allowed region to the left of the turning point t, (a) gives a schematic picture of the function $-Q^2(x)$, and (b) shows the phase of $Q^{1/2}(x)$.

where μ is the μ-integral (2.3.18) for the above-mentioned path joining x_1 and x_2. The connection problem involving connection from x_1 to x_2 or from x_2 to x_1 is displayed in the properties of $\mathbf{F}(x_1, x_2)$, and its solution is contained in (2.4.2a,b) and (2.4.3a,b).

2.4.2 Parameterization of the matrix $\mathbf{F}(x_1, x_2)$

To master complicated connection problems across a turning point efficiently it is convenient, as pointed out above, to parameterize the matrix $\mathbf{F}(x_1, x_2)$ as will be shown below.

With the aid of the symmetry relations (2.4.2a,b), the determinant relation (2.3.12a), i.e., $\det \mathbf{F}(x_1, x_2) = 1$, can be written

$$F_{21}(x_1, x_2)F_{11}^*(x_1, x_2) \mp i/2 = [F_{21}(x_1, x_2)F_{11}^*(x_1, x_2) \mp i/2]^*, \qquad (2.4.4)$$

and hence

$$F_{21}(x_1, x_2)F_{11}^*(x_1, x_2) \mp i/2 \text{ is a real quantity.} \qquad (2.4.5)$$

From (2.4.5), (2.4.2b) and (2.4.3a,b) one obtains the inequality

$$1 = \pm 2\,\mathrm{Im}[F_{21}(x_1, x_2)F_{11}^*(x_1, x_2)] = 2|\mathrm{Im}[F_{21}(x_1, x_2)F_{11}^*(x_1, x_2)]|$$
$$\leq 2|F_{21}(x_1, x_2)F_{11}^*(x_1, x_2)| \leq |\exp[2iw(x_1)]|(\mu + \text{higher powers of } \mu),$$

i.e.,

$$|\exp[-2iw(x_1)]| \lesssim \mu, \tag{2.4.6}$$

where in reality the quantity on the left-hand side is in general much smaller than μ. We define three real quantities α, β and γ as follows:

$$\alpha = \alpha(x_1, x_2) = |F_{11}(x_1, x_2)|, \tag{2.4.7a}$$

$$\beta = \beta(x_1, x_2) = \pm \arg F_{11}(x_1, x_2), \tag{2.4.7b}$$

where the upper and lower signs refer to the cases in Fig. 2.4.1 and Fig. 2.4.2, respectively, and

$$\gamma = \gamma(x_1, x_2) = [F_{21}(x_1, x_2) F_{11}^*(x_1, x_2) \mp i/2]/|F_{11}(x_1, x_2)|^2 \tag{2.4.7c}$$

$$= \frac{\mathrm{Re}[F_{21}(x_1, x_2) F_{11}^*(x_1, x_2)]}{|F_{11}(x_1, x_2)|^2} \tag{2.4.7c'}$$

$$= \mathrm{Re}[F_{21}(x_1, x_2)/F_{11}(x_1, x_2)]. \tag{2.4.7c''}$$

Here we have used (2.4.5) to obtain (2.4.7c'). From (2.4.7a,b,c''), the symmetry relation (2.4.2b) and the estimates (2.4.3a,b) it follows that

$$\alpha = 1 + O(\mu), \tag{2.4.8a}$$

$$\beta = O(\mu), \tag{2.4.8b}$$

$$\gamma = O(\mu) \exp[2iw(x_1)] = O(\mu) \exp[2\,|w(x_1)|], \tag{2.4.8c}$$

where the μ-integral is assumed to be small compared with unity, and $O(\mu)$ denotes quantities that are at the most of the order of magnitude μ.

With the aid of (2.4.7a–c) we can express $F_{11}(x_1, x_2)$ and $F_{21}(x_1, x_2)$ in terms of α, β and γ, and with the aid of (2.4.2a,b) we can then obtain expressions for $F_{12}(x_1, x_2)$ and $F_{22}(x_1, x_2)$. The result is

$$F_{11}(x_1, x_2) = \alpha \exp(\pm i\beta), \tag{2.4.9a}$$

$$F_{12}(x_1, x_2) = \pm i\alpha \exp(\mp i\beta), \tag{2.4.9b}$$

$$F_{21}(x_1, x_2) = \left(\alpha\gamma \pm \frac{i}{2\alpha} \right) \exp(\pm i\beta), \tag{2.4.9c}$$

$$F_{22}(x_1, x_2) = \left(\frac{1}{2\alpha} \pm i\alpha\gamma \right) \exp(\mp i\beta), \tag{2.4.9d}$$

where, as before, the upper and lower signs refer to the cases in Fig. 2.4.1 and 2.4.2, respectively. With α, β and γ defined by (2.4.7a–c), formulas (2.4.9a–d) are exact formulas, which are also valid if x_1 and x_2 lie close to the turning point. However, it is not easy to calculate α, β and γ exactly, and the great usefulness of our exact formulas is that by using the estimates (2.4.8a–c) we can in a straightforward and quite controllable way solve even complicated connection problems. We shall assume that x_1 and x_2 lie sufficiently far away from the well-isolated turning point t in order that the μ-integral in (2.4.8a–c) be small. The quantity α is then approximately equal to unity, the absolute value of β is small compared with unity, while γ is an unknown quantity, the absolute value of which in general is large compared with unity and difficult to determine more precisely than according to (2.4.8c).

2.4.2.1 Changes of α, β and γ when x_1 moves in the classically forbidden region

From (2.3.11) with (2.3.5b) we obtain, when the phases indicated in Figs. 2.4.1 and 2.4.2 are taken into account,

$$\frac{\partial}{\partial x_1} F_{11}(x_1, x_2) = \mp \frac{1}{2} \varepsilon(x_1) |q(x_1)| \{ F_{11}(x_1, x_2)$$
$$+ \exp[-2|w(x_1)|] F_{21}(x_1, x_2) \}, \qquad (2.4.10a)$$

$$\frac{\partial}{\partial x_1} F_{21}(x_1, x_2) = \pm \frac{1}{2} \varepsilon(x_1) |q(x_1)| \{ F_{21}(x_1, x_2)$$
$$+ \exp[2|w(x_1)|] F_{11}(x_1, x_2) \}, \qquad (2.4.10b)$$

where the upper and the lower signs refer to the cases in Figs. 2.4.1 and 2.4.2, respectively. Inserting into (2.4.10a) the expressions (2.4.9a,c) for $F_{11}(x_1, x_2)$ and $F_{21}(x_1, x_2)$, and identifying the real and imaginary terms on both sides of the resulting equation, we obtain, since $\varepsilon(x_1)$ is real,

$$\frac{\partial \ln \alpha}{\partial x_1} = \mp \frac{1}{2} \varepsilon(x_1) |q(x_1)| \{ 1 + \gamma \exp[-2|w(x_1)|] \}, \qquad (2.4.11a)$$

$$\frac{\partial \beta}{\partial x_1} = \mp \frac{1}{4\alpha^2} \varepsilon(x_1) |q(x_1)| \exp[-2|w(x_1)|]. \qquad (2.4.11b)$$

Inserting the expressions (2.4.9a,c) for $F_{11}(x_1, x_2)$ and $F_{21}(x_1, x_2)$ into (2.4.10b) and using (2.4.11a,b), we obtain

$$\frac{\partial \gamma}{\partial x_1} = \pm \frac{1}{2} \varepsilon(x_1) |q(x_1)| \left\{ \exp[+2|w(x_1)|] + 2\gamma + \left(\gamma^2 - \frac{1}{4\alpha^4} \right) \exp[-2|w(x_1)|] \right\}.$$
$$(2.4.11c)$$

We shall next obtain estimates for the changes of α, β and γ when x_1 moves to a point $\tilde{x}_1$ lying in the same classically forbidden region as the original point x_1.

From (2.4.11a–c) and (2.4.8a–c) we obtain

$$|\alpha(\tilde{x}_1, x_2) - \alpha(x_1, x_2)| \lesssim \frac{1}{2}\mu\,(x_1, \tilde{x}) + \text{higher powers of } \mu(x_1, \tilde{x}) \text{ and } \mu, \quad (2.4.12a)$$

$$|\beta(\tilde{x}_1, x_2) - \beta(x_1, x_2)| \lesssim \frac{1}{4}\mu(x_1, \tilde{x}_1)[1 + O(\mu)]\exp(-2K_1^{\min}), \quad (2.4.12b)$$

$$|\gamma\,(\tilde{x}_1, x_2) - \gamma\,(x_1, x_2)| \lesssim \frac{1}{2}\mu\,(x_1, \tilde{x}_1)\,[1 + O(\mu)]$$

$$\times\left[\exp(+2K_1^{\max}) + \frac{1}{4}\exp(-2K_1^{\min})\right]$$

$$\approx \frac{1}{2}\mu\,(x_1, \tilde{x}_1)\,[1 + O(\mu)]\exp(2K_1^{\max}), \quad (2.4.12c)$$

where μ is the maximum value of the μ-integral for an appropriate path in the upper half of the complex z-plane from x_2 to a point on the real axis between x_1 and $\tilde{x}_1$, $\mu(x_1, \tilde{x}_1)$ is the μ-integral

$$\mu(x_1, \tilde{x}_1) = \left|\int_{x_1}^{\tilde{x}_1} |\varepsilon(z)q(z)dz|\right| \quad (2.4.13)$$

along the real axis between x_1 and $\tilde{x}_1$, and $K_1^{\min}$ and $K_1^{\max}$ are the minimum and maximum, respectively, of the absolute value of $w(x)$, when x is an arbitrary point in the closed interval limited by x_1 and $\tilde{x}_1$. Since the phase of $q(x)$ is constant in that interval, we obviously have

$$K_1^{\min} = \text{Min}(|w(x_1)|, |w(\tilde{x}_1)|), \quad (2.4.14a)$$

$$K_1^{\max} = \text{Max}(|w(x_1)|, |w(\tilde{x}_1)|). \quad (2.4.14b)$$

2.4.2.2 Changes of α, β and γ when x_2 moves in the classically allowed region

From (2.3.11) with (2.3.5b) we obtain, when use is made of the inversion formula (2.3.12c) and the symmetry relations (2.4.2a,b) and the phases indicated in Fig. 2.4.1 and Fig. 2.4.2 are taken into account,

$$\frac{\partial}{\partial x_2}\ln F_{11}(x_1, x_2) = \frac{1}{2}\varepsilon(x_2)|q(x_2)|\{-i \mp \exp[\pm 2i\,|w(x_2)|]F_{11}^*(x_1, x_2)/F_{11}(x_1, x_2)\},$$

$$(2.4.15a)$$

$$\frac{\partial}{\partial x_2}F_{22}(x_1, x_2) = \frac{1}{2}\varepsilon(x_2)|q(x_2)|\{iF_{22}(x_1, x_2) \pm \exp[\mp 2i\,|w(x_2)|]F_{22}^*(x_1, x_2)\},$$

$$(2.4.15b)$$

where the upper and the lower signs refer to the cases in Figs. 2.4.1 and 2.4.2, respectively. Inserting into (2.4.15a) expression (2.4.9a) for $F_{11}(x_1, x_2)$, and identifying the real and imaginary terms on both sides of the resulting equation, we obtain, since $\varepsilon(x_2)$ is real,

$$\frac{\partial \ln \alpha(x_1, x_2)}{\partial x_2} = \mp \frac{1}{2} \varepsilon(x_2)|q(x_2)| \cos[2|w(x_2)| - 2\beta], \qquad (2.4.16a)$$

$$\frac{\partial \beta(x_1, x_2)}{\partial x_2} = \mp \frac{1}{2} \varepsilon(x_2)|q(x_2)|\{1 + \sin[2|w(x_2)| - 2\beta]\}. \quad (2.4.16b)$$

Inserting expression (2.4.9d) for $F_{22}(x_1, x_2)$ into (2.4.15b) and using (2.4.16a,b), we obtain

$$\frac{\partial \gamma(x_1, x_2)}{\partial x_2} = \mp \frac{1}{2\alpha^2} \varepsilon(x_2)|q(x_2)| \sin[2|w(x_2)| - 2\beta]. \qquad (2.4.16c)$$

It follows from (2.4.16a–c) and (2.4.8a) that

$$|\alpha(x_1, \tilde{x}_2) - \alpha(x_1, x_2)| \leq [1 + O(\mu)]\{\exp[\tfrac{1}{2}\mu(x_2, \tilde{x}_2)] - 1\}, \quad (2.4.17a)$$
$$|\beta(x_1, \tilde{x}_2) - \beta(x_1, x_2)| \leq \mu(x_2, \tilde{x}_2), \qquad (2.4.17b)$$
$$|\gamma(x_1, \tilde{x}_2) - \gamma(x_1, x_2)| \leq \tfrac{1}{2}[1 + O(\mu)]\mu(x_2, \tilde{x}_2), \qquad (2.4.17c)$$

where μ is the maximum value of the μ-integral for an appropriate path in the upper half of the complex z-plane from x_1 to a point on the real axis between x_2 and $\tilde{x}_2$, and, in analogy to (2.4.13), $\mu(x_2, \tilde{x}_2)$ is the μ-integral

$$\mu(x_2, \tilde{x}_2) = \left| \int_{x_2}^{\tilde{x}_2} |\varepsilon(z)q(z)dz| \right| \qquad (2.4.18)$$

along the real axis between x_2 and $\tilde{x}_2$. Although its absolute value is in general large, it is seen from (2.4.8c) that γ changes only very slightly when x_2 moves in the classically allowed region. We have achieved this result by introducing the square of $|F_{11}(x_1, x_2)|$ in the denominator of the definition (2.4.7c) of γ.

2.4.2.3 Limiting values of α, β and γ

We shall now consider the behaviour of α, β and γ as x_1 tends towards a certain limiting point X in such a way that the absolute value of $\exp[iw(x_1)]$ increases monotonically and tends to infinity, while the appropriate μ-integral is assumed to tend to a finite limiting value ($\ll 1$). The point X lies either at a finite position or is the limit of a point moving towards infinity. Using (2.3.12b) and noting that $iw(x_1) = |w(x_1)|$ according to Figs. 2.4.1 and 2.4.2, we find (with the aid of eqs. (4.5a–d) in Fröman and Fröman (1965)) that

$$\lim_{x_1 \to X} F_{11}(x_1, x_2) \text{ exists and is finite,} \tag{2.4.19a}$$

$$\lim_{x_1 \to X} F_{12}(x_1, x_2) \text{ exists and is finite,} \tag{2.4.19b}$$

$$\lim_{x_1 \to X} F_{21}(x_1, x_2) \exp[-2|w(x_1)|] = 0, \tag{2.4.19c}$$

$$\lim_{x_1 \to X} F_{22}(x_1, x_2) \exp[-2|w(x_1)|] = 0, \tag{2.4.19d}$$

where x_1 tends to X along the real axis. From the symmetry relations (2.4.2a,b) we see that (2.4.19b,d) follow from (2.4.19a,c). From (2.4.7a,b,c''), (2.4.8a,b) and (2.4.19a,c) it follows that

$$\lim_{x_1 \to X} \alpha(x_1, x_2) \text{ exists and is finite (close to unity),} \tag{2.4.20a}$$

$$\lim_{x_1 \to X} \beta(x_1, x_2) \text{ exists and is finite (close to zero),} \tag{2.4.20b}$$

$$\lim_{x_1 \to X} \gamma(x_1, x_2) \exp[-|2w(x_1)|] = 0. \tag{2.4.20c}$$

2.4.3 Wave function on opposite sides of a well-isolated turning point

In this section the exact wave function is expressed by means of the parameterized matrix $\mathbf{F}(x_1, x_2)$, i.e., in terms of α, β and γ, in such a form that the cumbersome work with different cases, depending on the choice of phase for $q(z)$, is eliminated. The practical work with the wave function is thus simplified considerably. The parameterized formulas for the wave function provide a tool for working easily with the exact solution in a straightforward and controllable way, such that in the final formulas one can clearly see which approximations can be made safely.

We still consider the two situations illustrated in Figs. 2.4.1 and 2.4.2 simultaneously. The difference between these two cases appears only in the signs of certain quantities, and as before we shall give both signs in the formulas, letting the upper and the lower signs refer to the cases in Fig. 2.4.1 and Fig. 2.4.2, respectively. Noting that with $w(z)$ defined by (2.4.1), where t is the turning point, we have according to (2.3.1a,b) along with Figs. 2.4.1 and 2.4.2

$$f_1(x_1) = \exp(\mp i\pi/4)|q^{-1/2}(x_1)| \exp[+|w(x_1)|], \tag{2.4.21a}$$

$$f_2(x_1) = \exp(\mp i\pi/4)|q^{-1/2}(x_1)| \exp[-|w(x_1)|]. \tag{2.4.21b}$$

Defining the quantities $C(x_1)$ and $D(x_1)$ by putting

$$a_1(x_1) = \exp(\pm i\pi/4)C(x_1), \quad a_2(x_1) = \exp(\pm i\pi/4)D(x_1), \tag{2.4.22a,b}$$

we obtain from (2.3.4a) and (2.4.21a,b) the wave function

$$\psi(x_1) = C(x_1)\left|q^{-1/2}(x_1)\right| \exp[+|w(x_1)|] + D(x_1)\left|q^{-1/2}(x_1)\right| \exp[-|w(x_1)|],$$
(2.4.23)

and according to (2.3.4b) we obtain the derivative $d\psi(x_1)/dx_1$ from (2.4.23) by treating $C(x_1)$ and $D(x_1)$ *formally* as if they were constants, i.e., independent of x_1. From (2.3.7) and the inversion formula (2.3.12c) it follows that

$$a_1(x_2) = F_{22}(x_1, x_2)a_1(x_1) - F_{12}(x_1, x_2)a_2(x_1),$$
(2.4.24a)
$$a_2(x_2) = -F_{21}(x_1, x_2)a_1(x_1) + F_{11}(x_1, x_2)a_2(x_1),$$
(2.4.24b)

and according to (2.3.1a,b) together with Figs. 2.4.1 and 2.4.2 we have

$$f_1(x_2) = \left|q^{-1/2}(x_2)\right| \exp[\pm i|w(x_2)|],$$
(2.4.25a)
$$f_2(x_2) = \left|q^{-1/2}(x_2)\right| \exp[\mp i|w(x_2)|].$$
(2.4.25b)

From (2.3.4a), (2.4.24a,b), (2.4.25a,b) and (2.4.22a,b) we obtain

$$\psi(x_2) = A(x_2)\left|q^{-1/2}(x_2)\right| \exp(+i|w(x_2)|) + B(x_2)\left|q^{-1/2}(x_2)\right| \exp(-i|w(x_2)|),$$
(2.4.26)

where

$$A(x_2) = \begin{cases} \exp(+i\pi/4)[F_{22}(x_1, x_2)C(x_1) - F_{12}(x_1, x_2)D(x_1)], & \text{Fig. 2.4.1,} \\ \exp(-i\pi/4)[-F_{21}(x_1, x_2)C(x_1) + F_{11}(x_1, x_2)D(x_1)], & \text{Fig. 2.4.2,} \end{cases}$$
(2.4.27a)

$$B(x_2) = \begin{cases} \exp(+i\pi/4)[-F_{21}(x_1, x_2)C(x_1) + F_{11}(x_1, x_2)D(x_1)], & \text{Fig. 2.4.1,} \\ \exp(-i\pi/4)[F_{22}(x_1, x_2)C(x_1) - F_{12}(x_1, x_2)D(x_1)], & \text{Fig. 2.4.2.} \end{cases}$$
(2.4.27b)

According to (2.3.4b) we obtain the derivative $d\psi(x_2)/dx_2$ from (2.4.26) by treating $A(x_2)$ and $B(x_2)$ *formally* as if they were constants, i.e., independent of x_2. Using (2.4.9a–d), we obtain from (2.4.27a,b) the formulas

$$A(x_2) = \left\{ \frac{1}{2\alpha}C(x_1) + i\alpha[\gamma C(x_1) - D(x_1)] \right\} \exp[+i(\pi/4 - \beta)],$$
(2.4.28a)

$$B(x_2) = \left\{ \frac{1}{2\alpha}C(x_1) - i\alpha[\gamma C(x_1) - D(x_1)] \right\} \exp[-i(\pi/4 - \beta)],$$
(2.4.28b)

which cover the situations in Figs. 2.4.1 and 2.4.2. Conversely, the coefficients

$C(x_1)$ and $D(x_1)$ can be expressed in terms of $A(x_2)$ and $B(x_2)$ by the formulas

$$C(x_1) = \alpha \exp[-i(\pi/4 - \beta)]A(x_2) + \alpha \exp[+i(\pi/4 - \beta)]B(x_2), \quad (2.4.29a)$$

$$D(x_1) = \left(\alpha\gamma + \frac{i}{2\alpha}\right) \exp[-i(\pi/4 - \beta)]A(x_2)$$

$$+ \left(\alpha\gamma - \frac{i}{2\alpha}\right) \exp[+i(\pi/4 - \beta)] B(x_2), \quad (2.4.29b)$$

which one obtains by solving (2.4.28a,b) with respect to $C(x_1)$ and $D(x_1)$. We note that if the wave function is chosen to be real on the real axis, $A(x_2)$ and $B(x_2)$ are complex conjugate, and $C(x_1)$ and $D(x_1)$ are real, which is obvious from (2.4.23) and (2.4.26). In expression (2.4.26) for $\psi(x_2)$ the coefficients $A(x_2)$ and $B(x_2)$ may be given *a priori*, or they may be determined in terms of $C(x_1)$ and $D(x_1)$ according to (2.4.28a,b). Analogously, in expression (2.4.23) for $\psi(x_1)$ the coefficients $C(x_1)$ and $D(x_1)$ may be given *a priori*, or they may be determined in terms of $A(x_2)$ and $B(x_2)$ according to (2.4.29a,b).

The wave function is given by (2.4.23) in the classically forbidden region and by (2.4.26) in the classically allowed region, and the coefficients in these formulas are related according to (2.4.28a,b) and (2.4.29a,b). In the classically allowed region the wave function can be written in an alternative form, if it is chosen to be real on the x-axis. To do this one puts

$$A(x_2) = \frac{1}{2}\Omega(x_2) \exp\{-i[\pi/4 - \delta(x_2)]\}, \quad (2.4.30a)$$

$$B(x_2) = \frac{1}{2}\Omega(x_2) \exp\{+i[\pi/4 - \delta(x_2)]\}, \quad (2.4.30b)$$

where $\Omega(x_2)$ is a positive amplitude, and $\delta(x_2)$ is a real phase. Inserting (2.4.30a,b) into (2.4.26), we obtain

$$\psi(x_2) = \Omega(x_2)\left|q^{-1/2}(x_2)\right| \cos[|w(x_2)| + \delta(x_2) - \pi/4]. \quad (2.4.31)$$

The formulas (2.4.23), (2.4.26), (2.4.28a,b), (2.4.29a,b), (2.4.30a,b) and (2.4.31) are *exact* formulas when α, β and γ are assumed to be given their exact values, and then the formulas in question are, of course, bi-directional. However, the precise values of α, β and γ are known only according to the estimates (2.4.8a–c), and therefore one obtains one-directional connection formulas.

2.4.4 Power and limitation of the parameterization method

The formulas obtained in this section may be very useful when there is a well-isolated turning point, i.e., a simple zero of $Q^2(z)$ on the real z-axis far away from other transition points. The parameterized formulas are in principle exact and thus bi-directional, but for the values of α, β and γ we have only the estimates (2.4.8a–c), (2.4.12a–c) and (2.4.17a–c). The procedure for using the formulas is quite straightforward and simple. One derives for the desired physical quantity an exact expression that contains the parameters α, β and γ associated with the relevant turning points. When, in this expression, the above-mentioned estimates are used, one can often, but not always, obtain for the desired physical quantity an approximate phase-integral expression in which the parameters α, β and γ no longer appear. As an example we mention that the rigorous treatment of the displacement of the energy levels due to the compression of an atom was a difficult problem, but with the aid of the parameterized formulas and the above-mentioned estimates of α, β and γ the treatment is straightforward and comparatively easy; see Section 3.29. As examples of situations in which the parameter γ remains in the resulting phase-integral formula we refer the reader to (3.10.13), (3.13.4a–c), (3.22.10a,b), Table 3.22.1, (3.22.17) and (3.23.9).

To illustrate further the power of the parameterized formulas we mention that by means of these formulas one can investigate the F-matrix connecting points in the classically forbidden regions on opposite sides of a broad single-well potential. It is found that this F-matrix has intricate and unexpected properties. The structure of the F-matrix in question in the general case is essentially the same as in the particular case of a purely parabolic potential well, which has been treated by Amaha (1993); we refer the reader to his paper.

As an example of the limitation of the parameterized formulas we mention that by means of these formulas, combined with the estimates (2.4.8a–c) of α, β and γ, one cannot derive, for a thick, real potential barrier, the approximate connection matrix that one obtains from (3.45.9a) in Section 3.45 by neglecting $\tilde{\phi}$. To achieve this result one must use the connection formula for a barrier, which one obtains by treating the two turning points as a 'cluster', i.e., by taking them into account simultaneously as in Section 2.5 and 3.44.

The formulas in this section are essential for the treatment of several problems in Chapter 3.

2.5 Phase-integral connection formulas for a real, smooth, single-hump potential barrier

The connection formula for a complex barrier is derived in Section 3.44. Here we restrict ourselves to considering a real barrier.

For energies far below the top of a potential barrier, the turning points, i.e., the two relevant zeros of $Q^2(z)$ on the real z-axis, are well separated, and, when the wave function is given on one side of the barrier, one can obtain an approximate expression for the wave function on the other side of the barrier with the aid of the connection formulas pertaining to each of the two turning points, unless the energy is close to a resonance energy. However, if for a thick barrier one is interested in obtaining the wave function very accurately, if the energy is close to a resonance energy, or if the barrier is thin or the energy lies close to or above the top of the barrier, the problem of determining the wave function on opposite sides of the barrier becomes delicate. For a full understanding of such problems one must resort to a more rigorous and accurate analysis in which both turning points are taken into account simultaneously.

When, for a superdense barrier, the energy increases, the turning points approach each other and coincide at the top of the barrier. On further increase of the energy, beyond the top of the barrier, there appear two complex conjugate transition zeros instead of the two turning points on the real z-axis. It is obvious that the connection formulas pertaining to a simple turning point are not applicable for energies in the neighbourhood of or above the top of the barrier. In the phase of the wave function, one has also to take into account a certain supplementary quantity $\tilde{\phi}$, which has been determined by the comparison equation technique under the assumption that the top of the barrier is approximately parabolic (Fröman, Fröman and Lundborg 1996).

Fröman and Fröman (1970) analysed rigorously and in great detail the connection problem for a real, smooth, single-hump potential barrier by using an exact, general solution on a convenient form and estimating certain quantities, expressed in terms of F-matrix elements, which were found to be small and could thus be neglected. Given a solution on one side of the barrier, they obtained (for subbarrier penetration as well as for superbarrier transmission) an expression (with error bounds) for the solution on the other side of the barrier. This expression was in terms of the phase-integral approximation of an arbitrary order, but with the restriction that the square $Q^2(z)$ of the base function was chosen to be equal to the coefficient function $R(z)$ in the Schrödinger-like differential equation. For a certain supplementary quantity, which in this book is denoted by $\tilde{\phi}$, and which is the same as the quantity -2σ in Fröman and Fröman (1970), an approximate expression corresponding only to the first-order approximation was known at that time. The results in this section go beyond those in Fröman and Fröman (1970) in the sense that the restriction on the choice of the base function is removed and the barrier connection formula obtained is written in a form that is independent of the choice of phase of the base function. Furthermore, for the supplementary quantity $\tilde{\phi}$ approximate analytical expressions can now be obtained up to the thirteenth order of the phase-integral approximation from formulas for ϕ in Fröman, Fröman and Lundborg (1996), which were derived by the comparison equation technique. Using these expressions for $\tilde{\phi}$ in higher orders of the phase-integral approximation, extremely accurate results have been obtained in several applications to physical problems, where barrier transmission occurs.

The connection formulas for a real potential barrier, which will be presented in this section, combined with the arbitrary-order connection formulas pertaining to a turning point (N. Fröman 1970), which are valid also when $Q^2(z)$ differs from $R(z)$, can be used for solving large classes of physical problems. One can, for instance, as in Section 3.46, obtain the quantization condition for a double-well potential, which was previously derived in another way by N. Fröman (1966c) and by N. Fröman and Myhrman (1970), and has proved also to be very accurate when the energy is close to the top of the barrier; see Fröman, Fröman, Myhrman and Paulsson (1972). We also mention the scattering problem for the radial Schrödinger equation with three turning points, although this is not treated in this book.

We summarize the results concerning barrier transmission that will be presented in this section as follows. For an arbitrary order of the phase-integral approximation, we give formulas for the wave function on both sides of a real, smooth, single-hump potential barrier of general shape, under the assumption that the second derivative of the potential is not too close to zero at the top of the barrier. In these formulas there appears a quantity $\tilde{\phi}$, which is particularly important when the energy is close to the top of the barrier, but which is also important for a thick barrier, if one wants results of great accuracy. In Subsection 2.5.1 the formulas are in principle exact, while in Subsection 2.5.2 the formulas are approximate. Resonance effects are discussed in Subsection 2.5.2.2.

2.5.1 Exact expressions for the wave function on both sides of the barrier

The presentation below is based on the detailed treatment of the transmission problem for a real potential barrier by Fröman and Fröman (1970). Although it was assumed there that $Q^2(z) = R(z)$, the results obtained are valid also when $Q^2(z) \neq R(z)$. In this book it is convenient partly to introduce notation other than that in Fröman and Fröman (1970). Thus the two relevant zeros of $Q^2(z)$ are now called t' and t'' (Re $t' \leq$ Re t'', Im $t' \leq 0$, Im $t'' \geq 0$). They are real turning points in the subbarrier case and complex conjugate transition zeros in the superbarrier case; see Fig. 3.45.1. Let x' be a point in the classically allowed region of the real z-axis to the left of the barrier and x'' be a point in the classically allowed region of the real z-axis to the right of the barrier. We introduce the notation

$$\theta = |F_{22}| \exp(K), \tag{2.5.1a}$$
$$\vartheta = \arg F_{22}, \tag{2.5.1b}$$
$$\tilde{\phi} = -2\sigma = \pi/2 - \arg F_{12}, \tag{2.5.1c}$$

where F_{12} and F_{22} are abbreviations for $F_{12}(x', x'')$ and $F_{22}(x', x'')$. We have also replaced σ in Fröman and Fröman (1970), by $-\tilde{\phi}/2$ in order to get better agreement with the notation used by other authors; see for instance eq. (C.50) in Child (1974). In the definitions (2.5.1a,b,c) it is assumed that the phase of $q^{1/2}(z)$ is chosen as in

Fröman and Fröman (1970), but in the following we write the formulas such that they are independent of the choice of phase for $Q^{1/2}(z)$ and hence of the phase for $q^{1/2}(z)$; see (2.2.11). The quantity K in (2.5.1a) is then given by

$$K = \frac{1}{2i} \int_\Lambda q(z)dz, \qquad (2.5.2)$$

where Λ is a closed contour of integration encircling both t' and t'', but no other transition point, and the integration is performed in the direction that in the *first-order approximation* yields $K > 0$ for energies below the top of the barrier and $K < 0$ for energies above the top of the barrier. If higher-order approximations are used, the quantity K may also be negative for energies somewhat below the top of the barrier, and hence the phase-integral formula (3.49.7) may yield values of the transmission coefficient larger than 1/2; see Table 3.49.1. Now we define

$$A' = B_1 \exp(+i\pi/4), \qquad (2.5.3a)$$
$$B' = A_1 \exp(-i\pi/4), \qquad (2.5.3b)$$

where the notation on the right-hand sides is that used by Fröman and Fröman (1970). Using the short-hand notation on the right-hand side of (2.2.15), we have, according to Fröman and Fröman (1970),

$$\psi(x') = A' \left| q^{-1/2}(x') \right| \exp \left[+i \left| \mathrm{Re} \int_{(t')}^{x'} q(z)dz \right| \right]$$

$$+ B' \left| q^{-1/2}(x') \right| \exp \left[-i \left| \mathrm{Re} \int_{(t')}^{x'} q(z)dz \right| \right], \qquad (2.5.4a)$$

$$\psi(x'') = A'' \left| q^{-1/2}(x'') \right| \exp \left[+i \left| \mathrm{Re} \int_{(t'')}^{x''} q(z)dz \right| \right]$$

$$+ B'' \left| q^{-1/2}(x'') \right| \exp \left[-i \left| \mathrm{Re} \int_{(t'')}^{x''} q(z)dz \right| \right], \qquad (2.5.4b)$$

where

$$\begin{pmatrix} A'' \\ B'' \end{pmatrix} = \tilde{\mathbf{M}} \begin{pmatrix} A' \\ B' \end{pmatrix} \qquad (2.5.5)$$

with

$$\tilde{\mathbf{M}} = \begin{pmatrix} \theta \exp[-i(\pi/2 + \vartheta)] & (\theta^2 + 1)^{1/2} \exp(+i\tilde{\phi}) \\ (\theta^2 + 1)^{1/2} \exp(-i\tilde{\phi}) & \theta \exp[+i(\pi/2 + \vartheta)] \end{pmatrix}, \qquad \det \tilde{\mathbf{M}} = -1.$$

$$\qquad (2.5.6a,b)$$

This matrix must not be confused with the matrix $\mathbf{M}$ defined in (2.3.5b). It is seen from (2.5.4a,b) that A' and A'' are associated with waves travelling in the direction away from the barrier, while B' and B'' are associated with waves coming in towards the barrier. From (2.5.6a,b) we get

$$\tilde{\mathbf{M}}^{-1} = \begin{pmatrix} \theta \exp[-i(\pi/2 - \vartheta)] & (\theta^2 + 1)^{1/2} \exp(+i\tilde{\phi}) \\ (\theta^2 + 1)^{1/2} \exp(-i\tilde{\phi}) & \theta \exp[+i(\pi/2 - \vartheta)] \end{pmatrix}, \qquad (2.5.7)$$

and hence one obtains $\tilde{\mathbf{M}}^{-1}$ from $\tilde{\mathbf{M}}$ by replacing ϑ with $-\vartheta$. From (2.5.5), (2.5.6a) and (2.5.7) it is seen that $\tilde{\phi}$ remains unchanged, whereas ϑ changes sign when one changes the connection from left to right to the connection from right to left. When the barrier is mirror symmetric, and the points x' and x'' lie mirror symmetrically with respect to the barrier, we have according to Fröman and Fröman (1970) the exact formula

$$\vartheta = 0 \text{ (mirror symmetric barrier)}, \qquad (2.5.8)$$

which is valid even if the barrier is not well isolated. According to what was said immediately below (2.5.7) we then have $\tilde{\mathbf{M}}^{-1} = \tilde{\mathbf{M}}$, and the matrix for connection in the direction from left to right is thus the same as the matrix for connection from right to left.

With A' and B' given (determined by a wave function known at the point x') the coefficients A'' and B'', which are obtained from (2.5.5) along with (2.5.6a), depend on x' and x'' via the quantities θ, ϑ and $\tilde{\phi}$. However, one obtains the derivatives of $\psi(x')$ and $\psi(x'')$ from (2.5.4a,b) by considering A', B', A'' and B'' *formally* as constants.

According to Fröman and Fröman (1970) (2.5.4a,b) along with (2.5.5) and (2.5.6a), and with θ, ϑ and $\tilde{\phi}$ defined by (2.5.1a–c), constitute *exact* expressions for $\psi(x')$ and $\psi(x'')$, whether or not the potential barrier is well separated from transition points that are not associated with the barrier.

2.5.2 Phase-integral connection formulas for a real barrier

Fröman and Fröman (1970) have shown that

$$\theta = \exp(K)\{1 + \exp[\tfrac{1}{2}(|K| - K)]O(\mu)\}, \qquad \vartheta = \exp[\tfrac{1}{2}(|K| - K)]O(\mu),$$
$$(2.5.9a,b)$$

where μ is the integral (2.3.18) along an appropriate path from x' to x'' avoiding t' and t''. When other transition points move away from t' and t'', so that the path of integration can also be moved farther away from t' and t'', the μ-integral diminishes and tends to zero in the limit of an exactly parabolic barrier if x' and x'' also move to infinity. The μ-integral also diminishes when one proceeds to higher orders of the phase-integral approximation, as long as the optimum order is not reached. When transition points that are not associated with the barrier lie sufficiently far away from t' and t'', the μ-integral in (2.5.9a,b) is so small that approximately

$$\theta \approx \exp(K), \quad \vartheta \approx 0. \tag{2.5.10a,b}$$

The quantity $\tilde{\phi}$ is particularly important when the energy is close to the top of the barrier, but it is also important for energies below the top if one wants very accurate results with the use of higher orders of the phase-integral approximation. In practice one cannot obtain useful expressions for $\tilde{\phi}$ from the exact formula (2.5.1c), but by using the comparison equation technique, and under the assumption that $d^2 Q^2(z)/dz^2$ is not too close to zero at the top of the barrier, Fröman, Fröman and Lundborg (1996) derived an approximate, but very accurate, formula in the $(2N+1)$th order of the phase-integral approximation for the closely related quantity ϕ associated with a complex barrier; see their eqs. (5.5.30) and (5.5.25a–g). For a real barrier ϕ is real and equal to $\tilde{\phi}$ when the barrier is superdense, while ϕ is complex (with the real part equal to $\tilde{\phi}$) when the barrier is underdense; see Section 3.45. According to Fröman, Fröman and Lundborg (1996) (eq. (5.5.30) with $\lambda = 1$) one therefore has

$$\tilde{\phi} = \arg\Gamma(1/2 + i\bar{K}) - \bar{K}\ln|\bar{K}_0| + \sum_{n=0}^{N} \phi^{(2n+1)}, \tag{2.5.11}$$

where $\bar{K}$ and $\bar{K}_0$ are obtained from eqs. (5.4.23) and (5.4.21) in Fröman, Fröman and Lundborg (1996), i.e.,

$$\bar{K} = \sum_{n=0}^{N} \bar{K}_{2n} = K/\pi, \tag{2.5.12a}$$

$$\bar{K}_{2n} = \frac{1}{2\pi i} \int_{\Lambda} Y_{2n} Q(z)dz, \quad n = 0, 1, 2, \ldots, \tag{2.5.12b}$$

Λ being the previously described contour of integration encircling t' and t'', but no other transition point, with the integration performed in the direction that makes $\bar{K}_0$ positive when t' and t'' are real, i.e., when the barrier is superdense, but negative when t' and t'' are complex conjugate, i.e., when the barrier is underdense, and $\phi^{(1)}, \ldots, \phi^{(13)}$ are given by eqs. (5.5.25a–g) in Fröman, Fröman and Lundborg (1996). Here we quote the expressions for the first three of these quantities:

$$\phi^{(1)} = \bar{K}_0, \tag{2.5.13a}$$

$$\phi^{(3)} = -\frac{1}{24\bar{K}_0}, \tag{2.5.13b}$$

$$\phi^{(5)} = -\frac{\bar{K}_2^2}{2\bar{K}_0} + \frac{\bar{K}_2}{24\bar{K}_0^2} - \frac{7}{2880\bar{K}_0^3}, \tag{2.5.13c}$$

where the quantities $\bar{K}_{2n}, n = 0, 1, 2, \ldots$, are defined by (2.5.12b). Formula (2.5.11) for $\tilde{\phi}$ along with (2.5.12a,b) and (2.5.13a–c) is valid for all values of the energy, even those close to the top of the barrier. Although (2.5.11) is also valid when the barrier is thick, it may be useful to have a simple, though more approximate, formula for this case. Such a formula for ϕ has been derived by Fröman, Fröman and Lundborg (1996) (their eq. (5.5.31)). Since for subbarrier penetration $\tilde{\phi} = \phi$, we have according to that formula in the $(2N + 1)$th order of approximation

$$\tilde{\phi} = -\phi^{(2N+3)} \text{ (thick barrier)}. \qquad (2.5.14)$$

From (2.5.14) along with (2.5.13b,c) it is seen that the absolute value of $\tilde{\phi}$ is small compared with unity when the barrier is thick.

   We emphasize again that for the validity of (2.5.11) with the expressions (2.5.13a–c) for $\phi^{(2n+1)}$ the essential restriction is that the absolute value of $d^2 Q^2(z)/dz^2$ must not be too small at the top of the barrier, which means that close to its top the barrier is approximately parabolic. However, when the energy is close to the top of the barrier, it is the slight deviation from parabolic shape close to the top of the barrier that determines the values of the quantities $\bar{K}_{2n}$, $n > 0$, and one needs accurate values of these quantities to obtain accurate values of $\tilde{\phi}$ in higher orders of the phase-integral approximation.

   In the work by Ford, Hill, Wakano and Wheeler (1959) on quantum effects near the top of a barrier that is parabolic over an appreciable region near its maximum there appears an expression that is equivalent to the first-order expression for $\tilde{\phi}$ given by (2.5.11) for $N = 0$ and (2.5.13a). Equivalent expressions for $\tilde{\phi}$, pertaining to the first-order approximation and the particular case of an exactly parabolic barrier, have also been derived by Soop (1965) and Connor (1968). These first-order expressions for $\tilde{\phi}$ for an exactly parabolic barrier have also been used when the barrier has only an approximately parabolic shape.

   Using (2.5.10a,b), we find from (2.5.4a,b), (2.5.5) and (2.5.6a) that if the wave function at the point x' to the left of the barrier is

$$\psi(x') = A' \left| q^{-1/2}(x') \right| \exp\left[+i \left| \mathrm{Re} \int_{(t')}^{x'} q(z) dz \right| \right]$$

$$+ B' \left| q^{-1/2}(x') \right| \exp\left[-i \left| \mathrm{Re} \int_{(t')}^{x'} q(z) dz \right| \right], \qquad (2.5.15a)$$

the same wave function at the point x'' to the right of the barrier is approximately

$$\psi(x'') = \exp(K)\{-iA' + [1 + \exp(-2K)]^{1/2}\exp(+i\tilde{\phi})B'\}$$

$$\times \left|q^{-1/2}(x'')\right|\exp\left[+i\left|\mathrm{Re}\int_{(t'')}^{x''} q(z)dz\right|\right]$$

$$+ \exp(K)\{[1 + \exp(-2K)]^{1/2}\exp(-i\tilde{\phi})A' + iB'\}$$

$$\times \left|q^{-1/2}(x'')\right|\exp\left[-i\left|\mathrm{Re}\int_{(t'')}^{x''} q(z)dz\right|\right]. \qquad (2.5.15b)$$

Since the expressions for the coefficients of the phase-integral functions in this formula are approximate, the value of a certain coefficient is not reliable if its absolute value is too small compared with the sum of the absolute values of the two terms in the expression for the coefficient.

In the following two subsections we shall discuss the two important particular cases in which, to the left of the barrier, the wave function is given either as an outgoing wave or as a standing wave. The treatment of the latter case, which leads to a discussion of resonance phenomena, shows that when certain precautions are taken our approximate solution accounts in a very precise way for the narrow resonances that are of special interest in scattering and decay problems.

2.5.2.1 Wave function given as an outgoing wave to the left of the barrier

When the wave function represents an outgoing wave to the left of the barrier, we put $A' = 1$ and $B' = 0$. From (2.5.15a,b) we then obtain

$$\psi(x') = \left|q^{-1/2}(x')\right|\exp\left[+i\left|\mathrm{Re}\int_{(t')}^{x'} q(z)dz\right|\right], \qquad (2.5.16a)$$

$$\psi(x'') = -i\exp(K)\left|q^{-1/2}(x'')\right|\exp\left[+i\left|\mathrm{Re}\int_{(t'')}^{x''} q(z)dz\right|\right]$$

$$+ [\exp(2K) + 1]^{1/2}\exp(-i\tilde{\phi})\left|q^{-1/2}(x'')\right|\exp\left[-i\left|\mathrm{Re}\int_{(t'')}^{x''} q(z)dz\right|\right]. $$

$$(2.5.16b)$$

As already mentioned, the quantity $\tilde{\phi}$ is particularly important for the phase of the wave function at the point x'' when the energy is close to the top of the barrier. However, when the phase of the wave function is irrelevant, $\tilde{\phi}$ is not important.

Thus, the transmission coefficient T, which can easily be derived from (2.5.16a,b), is given for all energies by the simple formula (see Section 3.49)

$$T = \frac{1}{(\theta^2 + 1)^2} \approx \frac{1}{\exp(2K) + 1}. \qquad (2.5.17)$$

When $\exp(K) \gg 1$, i.e., for energies well below the top of the barrier, the connection formulas pertaining to the separate turning points yield the less accurate formula $T = \exp(-2K)$.

As already mentioned, in higher-order approximations K may also become negative for energies somewhat below the top of the barrier, and the transmission coefficient may therefore be obtained approximately from (2.5.17) with the aid of higher-order approximations for steep and thin barriers, when the first-order approximation is not applicable at all. See Table 3.49.1 in Section 3.49.

2.5.2.2 Wave function given as a standing wave to the left of the barrier

The case in which the wave function is given as a standing wave on one side of the barrier requires a detailed treatment, since the resonance phenomenon may occur. Putting in (2.5.4a)

$$A' = \tfrac{1}{2}\Omega' \exp[i(\delta' - \pi/4)], \qquad (2.5.18a)$$

$$B' = \tfrac{1}{2}\Omega' \exp[-i(\delta' - \pi/4)], \qquad (2.5.18b)$$

where Ω' is a positive amplitude, and δ' is an arbitrary real phase constant, we get

$$\psi(x') = \Omega' \left| q^{-1/2}(x') \right| \cos \left[\left| \mathrm{Re} \int_{(t')}^{x'} q(z)dz \right| + \delta' - \pi/4 \right], \qquad (2.5.19a)$$

and putting in (2.5.4b)

$$A'' = \tfrac{1}{2}\Omega'' \exp[i(\delta'' - \pi/4)], \qquad (2.5.20a)$$

$$B'' = \tfrac{1}{2}\Omega'' \exp[-i(\delta'' - \pi/4)], \qquad (2.5.20b)$$

where Ω'' is a positive amplitude, and δ'' is a real phase constant, we obtain

$$\psi(x'') = \Omega'' \left| q^{-1/2}(x'') \right| \cos \left[\left| \mathrm{Re} \int_{(t'')}^{x''} q(z)dz \right| + \delta'' - \pi/4 \right]. \qquad (2.5.19b)$$

In order to be able to discuss resonance effects we need exact formulas admitting a detailed analysis of δ'' and Ω''. Inserting (2.5.18a,b) and (2.5.20a,b) into (2.5.5) and recalling (2.5.6a), we obtain

$$\Omega'' \exp[i(\delta'' - \tilde{\phi}/2 + \vartheta/2)] = \Omega'\{(\theta^2 + 1)^{1/2} \exp[+i(\pi/2 + \tilde{\phi}/2 + \vartheta/2 - \delta')] \\ + \theta \exp[-i(\pi/2 + \tilde{\phi}/2 + \vartheta/2 - \delta')]\} \quad (2.5.21)$$

and the complex conjugate of (2.5.21). Separating (2.5.21) into real and imaginary parts, we get

$$\Omega'' \cos(\delta'' - \tilde{\phi}/2 + \vartheta/2) = \Omega'\left[(\theta^2 + 1)^{1/2} + \theta\right] \cos(\pi/2 + \tilde{\phi}/2 + \vartheta/2 - \delta'),$$
$$(2.5.22a)$$

$$\Omega'' \sin(\delta'' - \tilde{\phi}/2 + \vartheta/2) = \Omega'\left[(\theta^2 + 1)^{1/2} - \theta\right] \sin(\pi/2 + \tilde{\phi}/2 + \vartheta/2 - \delta').$$
$$(2.5.22b)$$

According to (2.5.22a,b) the angle $\delta'' - \tilde{\phi}/2 + \vartheta/2$ lies in the same quadrant as the angle $\pi/2 + \tilde{\phi}/2 + \vartheta/2 - \delta'$ (mod 2π). From (2.5.22a,b) we get

$$\delta'' = \arctan\left[\frac{(\theta^2 + 1)^{1/2} - \theta}{(\theta^2 + 1)^{1/2} + \theta} \tan(\pi/2 + \tilde{\phi}/2 + \vartheta/2 - \delta')\right] + \tilde{\phi}/2 - \vartheta/2,$$
$$(2.5.23a)$$

$$\Omega'' = \Omega'\{[(\theta^2 + 1)^{1/2} - \theta]^2 + 4\theta(\theta^2 + 1)^{1/2} \cos^2(\pi/2 + \tilde{\phi}/2 + \vartheta/2 - \delta')\}^{1/2},$$
$$(2.5.23b)$$

where the branch of arctan is to be chosen such that δ'' is a continuous function of δ', and $\delta'' - \tilde{\phi}/2 + \vartheta/2$ lies in the same quadrant as $\pi/2 + \tilde{\phi}/2 + \vartheta/2 - \delta'$ (mod 2π). Alternatively we can write (2.5.23a) as

$$\tan(\delta' - \tilde{\phi}/2 - \vartheta/2)\tan(\delta'' - \tilde{\phi}/2 + \vartheta/2) = \frac{(\theta^2 + 1)^{1/2} - \theta}{(\theta^2 + 1)^{1/2} + \theta}. \quad (2.5.23a')$$

From (2.5.23b) it follows that

$$(\theta^2 + 1)^{1/2} - \theta \le \Omega''/\Omega' \le (\theta^2 + 1)^{1/2} + \theta, \quad (2.5.24)$$

where the equality sign to the left is valid when $\delta' - \tilde{\phi}/2 - \vartheta/2$ is an integer multiple of π, and the equality sign to the right is valid when $\delta' - \tilde{\phi}/2 - \vartheta/2 - \pi/2$ is an integer multiple of π.

From (2.5.23a,b) it is seen that $\delta'' - \tilde{\phi}/2 + \vartheta/2$ is a monotonically decreasing function of $\delta' - \tilde{\phi}/2 - \vartheta/2$, whereas Ω''/Ω' attains maxima and minima when $\delta' - \tilde{\phi}/2 - \vartheta/2$ changes. Under the assumption that the barrier is thick ($\theta \gg 1$), we present in more detail in Table 2.5.1 the dependence of $\delta'' - \tilde{\phi}/2 + \vartheta/2$ and Ω''/Ω' on $\delta' - \tilde{\phi}/2 - \vartheta/2$. It is seen that for thick barriers δ'' is close to an integer multiple of π, except when δ' lies in certain intervals of the order of magnitude $1/\theta^2$.

Table 2.5.1. *Change of phase and amplitude of the wave function on one side of a barrier, due to change of phase on the other side.*

$\delta' - \tilde{\phi}/2 - \vartheta/2$	$\delta'' - \tilde{\phi}/2 + \vartheta/2$	Ω''/Ω'
$-\pi/2$	π (exact)	2θ (maximum)
$-\pi/4$	$\pi - 1/(4\theta^2)$	$2^{1/2}\theta$
$-1/(2\theta)$	$\pi - 1/(2\theta)$	1
$-1/(4\theta^2)$	$3\pi/4$	$1/(2^{1/2}\theta)$
0	$\pi/2$ (exact)	$1/(2\theta)$ (minimum)
$1/(4\theta^2)$	$\pi/4$	$1/(2^{1/2}\theta)$
$1/(2\theta)$	$1/(2\theta)$	1
$\pi/4$	$1/(4\theta^2)$	$2^{1/2}\theta$
$\pi/2$	0 (exact)	2θ (maximum)
$3\pi/4$	$-1/(4\theta^2)$	$2^{1/2}\theta$
$\pi - 1/(2\theta)$	$-1/(2\theta)$	1
$\pi - 1/(4\theta^2)$	$-\pi/4$	$1/(2^{1/2}\theta)$
π	$-\pi/2$ (exact)	$1/(2\theta)$ (minimum)
$\pi + 1/(4\theta^2)$	$-3\pi/4$	$1/(2^{1/2}\theta)$
$\pi + 1/(2\theta)$	$-\pi + 1/(2\theta)$	1
$5\pi/4$	$-\pi + 1/(4\theta^2)$	$2^{1/2}\theta$
$3\pi/2$	$-\pi$ (exact)	2θ (maximum)

In these small intervals δ'' changes rapidly between values close to neighbouring multiples of π, while Ω''/Ω' is of the order of magnitude $1/\theta$. When δ' moves to the left or to the right of such a small interval, δ'' remains approximately constant and is close to an integer multiple of π, while Ω''/Ω' increases up to maximum values approximately equal to 2θ.

From the connection formulas (3.10.10') and (3.12.9) pertaining to well-isolated turning points it follows that $\sin \delta''$ is close to zero and Ω''/Ω' is large compared with unity unless $\sin \delta'$ is close to zero, while $\sin \delta''$ is significantly different from zero and Ω''/Ω' is small compared to unity in a narrow range of values for $\sin \delta'$ close to zero. These qualitative results are in agreement with the more detailed results obtained from (2.5.23a,b) and presented in Table 2.5.1.

Now we rewrite (2.5.21) as

$$\Omega''/\Omega' = i(\theta^2 + 1)^{1/2} \exp[-i(\delta' + \delta'' - \tilde{\phi})] - i\theta \exp[i(\delta' - \delta'' - \vartheta)]$$
$$= (\theta^2 + 1)^{1/2} \sin(\delta' + \delta'' - \tilde{\phi}) + \theta \sin(\delta' - \delta'' - \vartheta)$$
$$+ i[(\theta^2 + 1)^{1/2} \cos(\delta' + \delta'' - \tilde{\phi}) - \theta \cos(\delta' - \delta'' - \vartheta)]. \quad (2.5.25)$$

Since Ω''/Ω' is real and positive, we obtain from (2.5.25) the two formulas

$$\cos(\delta' + \delta'' - \tilde{\phi}) = \frac{\theta}{(\theta^2 + 1)^{1/2}} \cos(\delta' - \delta'' - \vartheta), \quad (2.5.26)$$

$$\Omega''/\Omega' = (\theta^2 + 1)^{1/2} \sin(\delta' + \delta'' - \tilde{\phi}) + \theta \sin(\delta' - \delta'' - \vartheta). \quad (2.5.27)$$

From (2.5.26) it follows that

$$|\sin(\delta' + \delta'' - \tilde{\phi})| = \left[\frac{1 + \theta^2 \sin^2(\delta' - \delta'' - \vartheta)}{(\theta^2 + 1)}\right]^{1/2}. \qquad (2.5.28)$$

Therefore

$$(\theta^2 + 1)^{1/2}|\sin(\delta' + \delta'' - \tilde{\phi})| > \theta|\sin(\delta' - \delta'' - \vartheta)|. \qquad (2.5.29)$$

Since Ω''/Ω' is positive, it follows from (2.5.27) and (2.5.29) that $\sin(\delta' + \delta'' - \tilde{\phi})$ is positive, and from (2.5.28) we therefore obtain

$$\sin(\delta' + \delta'' - \tilde{\phi}) = \left[\frac{1 + \theta^2 \sin^2(\delta' - \delta'' - \vartheta)}{\theta^2 + 1}\right]^{1/2}. \qquad (2.5.30)$$

Inserting (2.5.30) into (2.5.27), we get

$$\Omega''/\Omega' = [1 + \theta^2 \sin^2(\delta' - \delta'' - \vartheta)]^{1/2} + \theta \sin(\delta' - \delta'' - \vartheta). \qquad (2.5.31)$$

As long as the exact expressions (2.5.1a–c) for θ, ϑ and $\tilde{\phi}$ are retained, the wave function given by (2.5.19a) at x' is exactly given by (2.5.19b), with Ω'' and δ'' obtained from (2.5.23a,b), at x'' on the other side of the barrier. We now introduce the approximate value $\theta = \exp(K)$ according to (2.5.10a) and rewrite (2.5.23a,b) in a form that displays the resonance structure of $\psi(x'')$:

$$\delta'' = \arctan\left\{\frac{[1 + \exp(-2K)]^{1/4} - [1 + \exp(-2K)]^{-1/4}}{[1 + \exp(-2K)]^{1/4} + [1 + \exp(-2K)]^{-1/4}}\right.$$
$$\left. \times \tan(\pi/2 + \tilde{\phi}/2 + \vartheta/2 - \delta')\right\} + \tilde{\phi}/2 - \vartheta/2 \qquad (2.5.32a)$$

$$\frac{1}{\Omega''} = \frac{\exp(K)\{[1 + \exp(-2K)]^{1/2} + 1\}}{\Omega'\left(1 + \left\{\dfrac{\sin(\delta' - \tilde{\phi}/2 - \vartheta/2)}{\frac{1}{2}[1 + \exp(-2K)]^{1/4} - \frac{1}{2}[1 + \exp(-2K)]^{-1/4}}\right\}^2\right)^{1/2}}, \qquad (2.5.32b)$$

where arctan is to be chosen such that $\delta'' - \tilde{\phi}/2 + \vartheta/2$ lies in the same quadrant as $\pi/2 + \tilde{\phi}/2 + \vartheta/2 - \delta'$(mod 2π). According to (2.5.9b) the absolute value of ϑ is expected to be small compared with unity unless the energy lies too far above the top of the barrier, and ϑ decreases rapidly when one proceeds from the first order to higher orders of the phase-integral approximation. In practice one has in general to omit the quantity ϑ, since practically useful formulas for ϑ, corresponding to (2.5.11) along with (2.5.12a,b) and (2.5.13a–c), have not been obtained. However, in the particular case when the barrier is mirror symmetric and x' and x'' lie mirror symmetrically with respect to the barrier, ϑ is known to be exactly equal to zero according to (2.5.8).

We shall now consider the two special situations one in which the energy lies in the neighbourhood of the top of the barrier and another in which the energy lies well below the top of the barrier.

For a superdense barrier, when the energy increases and approaches the top of the barrier, the maxima of $1/\Omega''$ become increasingly broadened. Thus, for energies close to the top of the barrier, i.e., when $\exp(2K)$ is of the order of unity, the approximation $\vartheta \approx 0$ in (2.5.32b) is not dangerous, since it gives rise to displacements in the positions of the maxima that are small in comparison with their half-widths.

For energies well below the top of the barrier, i.e., when $\exp(-2K) \ll 1$, the quantity $1/\Omega''$ has according to (2.5.32b) pronounced maxima when $\delta' - \tilde{\phi}/2 - \vartheta/2$ is an integer multiple of π. If x' lies in the interior of some atomic system, and x'' lies outside this system, the maxima of $1/\Omega''$ correspond to resonances for the physical system. Quasi-stationary energy levels of a hydrogen atom in an electric field and orbiting resonances in molecular scattering provide examples of this type of resonance. As $1/\Omega''$ passes through a maximum, when δ' increases, the value of the phase $\delta'' - \tilde{\phi}/2 + \vartheta/2$ changes by $-\pi$ according to (2.5.32a). The resonances are very sharp with half-widths approximately equal to $\frac{1}{4}\exp(-2K)$ according to (2.5.32b). Thus, although the quantity $\tilde{\phi}/2 + \vartheta/2$ is very small well below the top of the barrier, it is still of decisive importance close to a sharp resonance, since, due to the largeness of the exponential $\exp(K)$, the omission of $\tilde{\phi}/2 + \vartheta/2$ in (2.5.32a,b), or even the omission of only $\vartheta/2$, will in general imply that the resulting formulas for δ'' and $1/\Omega''$ will yield a displacement of the resonance from its actual position by powers of ten times the half-width. This has been confirmed by Lundborg (1977) for an exactly soluble model. If higher orders of the phase-integral approximation are used, the absolute value of ϑ decreases, and the absolute value of $\tilde{\phi}$ also decreases, as is seen from (2.5.14) and (2.5.13b,c), and by proceeding to conveniently high orders, one may achieve the result that the true position, even of a narrow resonance, is obtained to within a half-width, if one neglects ϑ, and sometimes if $\tilde{\phi}$ is also neglected. This fact was verified for an exactly soluble model by Lundborg (1977), and it is also illustrated by the extremely accurate results obtained for the positions and widths of the Stark levels for a hydrogen atom in an electric field (Fröman and Fröman 1984). It should be noted, however, that, since $\tilde{\phi}/2 + \vartheta/2$ is small and varies slowly with energy, these shifts of the resonances may not be essential in some contexts, e.g. in decay problems when one integrates with respect to the energy.

The formulas in this section are exact when θ, ϑ and $\tilde{\phi}$ are given their exact values, but when one introduces the approximations (2.5.10a,b) and (2.5.11) along with (2.5.12a,b) and (2.5.13a,b), one obtains approximate phase-integral formulas. These approximate formulas can also be obtained by the particularization in Section 3.45 to a real barrier of formulas for a complex barrier derived in Section 3.44. They will be used in Sections 3.46 and 3.49.

3

Problems with solutions

The problems in this chapter comprise the following items:

determination of a convenient base function (Sections 3.1 and 3.2);
determination of a phase-integral function satisfying the Schrödinger equation exactly (Section 3.3);
properties of the phase-integral approximation along certain paths (Sections 3.4–3.6);
Stokes constants and connection formulas (Sections 3.7–3.16);
Airy's differential equation (Sections 3.17–3.20);
change of phase of the wave equation in a classically allowed region due to the change of a boundary condition imposed in an adjacent classically forbidden region (Sections 3.21–3.24);
phase shift (Section 3.25);
near-lying energy levels (Section 3.26);
quantization conditions (Sections 3.27–3.32);
determination of the potential from the energy spectrum (Sections 3.33–3.35);
formulas for the normalization integral, not involving the wave function (Sections 3.36 and 3.37);
potential with a strong attractive Coulomb singularity at the origin (Sections 3.38 and 3.39);
formulas for expectation values and matrix elements, not involving the wave functions (Sections 3.40–3.43);
potential barriers (Sections 3.44–3.50).

3.1 Base function for the radial Schrödinger equation when the physical potential has at the most a Coulomb singularity at the origin

With the physical potential given as prescribed, and with the radial Schrödinger equation given by (2.1.1), we write $R(z)$ as

$$R(z) = -\frac{l(l+1)}{z^2} + \frac{a_{-1}}{z} + a_0 + a_1 z + \cdots, \qquad (3.1.1)$$

where l is the angular momentum quantum number, and $a_{-1}, a_0, a_1, \ldots$ are constants. Assume that the square of the base function is

$$Q^2(z) = \frac{b_{-2}}{z^2} + \frac{b_{-1}}{z} + b_0 + b_1 z + \cdots, \qquad (3.1.2)$$

where $b_{-2}, b_{-1}, b_0, b_1, \ldots$ are constants, and discuss, by considering the μ-integral (2.3.18) as z approaches the origin, the possibility of choosing the coefficients in (3.1.2) such that the first-order phase-integral approximation is valid close to the origin.

Solution. In order that the first-order approximation be valid close to the origin, we require that the integral $\mu(z, z_0)$, defined by (2.3.18) and particularized to the first-order approximation, be finite when $z \to 0$. In this approximation the integrand in the right-hand member of (2.3.18) is $|\varepsilon_0 Q|$ with ε_0 given by (2.2.1′). If $b_{-2} \neq 0$, straightforward calculations yield

$$\frac{1}{16Q^6}\left[5\left(\frac{dQ^2}{dz}\right)^2 - 4Q^2 \frac{d^2Q^2}{dz^2}\right] = -\frac{1}{4b_{-2}} + O(z^2), \qquad (3.1.3)$$

$$\frac{R - Q^2}{Q^2} = -\frac{l(l+1) + b_{-2}}{b_{-2}} + \left\{\frac{[l(l+1) + b_{-2}]b_{-1}}{b_{-2}^2} + \frac{a_{-1} - b_{-1}}{b_{-2}}\right\}z + O(z^2), \qquad (3.1.4)$$

and hence we obtain from (2.2.1′)

$$\varepsilon_0 = -\frac{(l + 1/2)^2 + b_{-2}}{b_{-2}} + \left\{\frac{[l(l+1) + b_{-2}]b_{-1}}{b_{-2}^2} + \frac{a_{-1} - b_{-1}}{b_{-2}}\right\}z + O(z^2), \qquad (3.1.5)$$

and thus

$$\varepsilon_0 Q = -\frac{(l + 1/2)^2 + b_{-2}}{b_{-2}^{1/2} z} + \frac{[(l + 1/2)^2 + b_{-2} - 1/2]b_{-1}}{2b_{-2}^{3/2}} + \frac{a_{-1} - b_{-1}}{b_{-2}^{1/2}} + O(z). \qquad (3.1.6)$$

According to (2.3.18) and (3.1.6) the requirement that for the first-order approximation $\mu(z, z_0)$ be finite when $z \to 0$ implies that

$$b_{-2} = -(l + 1/2)^2, \qquad (3.1.7)$$

i.e., according to (3.1.2)

$$\lim_{z \to 0} z^2 Q^2(z) = -(l + 1/2)^2, \qquad (3.1.8)$$

and according to (3.1.1), (3.1.2) and (3.1.7)

$$\lim_{z \to 0} z^2[Q^2(z) - R(z)] = -1/4. \tag{3.1.9}$$

We note that the integral $\mu(z, z_0)$ remains finite as $z \to 0$ independently of the values of b_v for $v \geq -1$. The most common choice of $Q^2(z)$ for making the wave function, in any order of approximation, valid at the origin is

$$Q^2(z) = R(z) - 1/(4z^2), \tag{3.1.10}$$

and with this choice the phase shift is exactly equal to zero for a free particle, and when $l \neq -1/2$ the wave function has the correct dependence on z in the neighbourhood of the origin, while these results are in general not achieved if one chooses $Q^2(z) = R(z)$.

One can also obtain the result (3.1.10) by means of the Langer transformation, which is described in many textbooks. In this connection it should be remarked that some authors motivate the Langer transformation by saying that the WKB approximation applies to the interval $(-\infty, +\infty)$, but not to the interval $(0, +\infty)$. This statement is obviously inadequate, since the extent of the interval is not involved in the derivation of the approximation.

3.2 Base function and wave function close to the origin when the physical potential is repulsive and strongly singular at the origin

Determine an appropriate base function and an approximate expression for the wave function in the neighbourhood of the origin when the effective potential behaves as c^2/z^m as $z \to 0$, where c is a positive constant and $m > 2$.

Solution. With

$$Q^2(z) = -c^2 z^{-m} + O(z^{1-m}) \quad \text{as } z \to 0, \quad c > 0, \tag{3.2.1}$$

we obtain from (2.2.1)

$$\varepsilon_0 = O(z^{m-2}) + [R(z) - Q^2(z)]O(z^m) \tag{3.2.2}$$

and hence

$$\varepsilon_0 Q = O(z^{m/2-2}) + [R(z) - Q^2(z)]O(z^{m/2}). \tag{3.2.3}$$

From (2.3.18) and (3.2.3) it is seen that the first-order integral $\mu(z, z_0)$ remains finite as z tends to zero, and that the regular wave function thus behaves correctly close to the origin if $R(z) - Q^2(z) = o(z^{-m/2-1})$. In particular one can thus choose either

$Q^2(z) = R(z)$ or $Q^2(z) = R(z) - 1/(4z^2)$. The latter choice may be preferred in phase-shift problems involving a single turning point, since one then obtains that the radial phase shift is exactly equal to zero for a free particle, i.e., when the physical potential is equal to zero; see Section 3.25.

In the classically forbidden region of the real axis the first-order phase-integral approximation gives, except for an arbitrary constant factor, for the physically acceptable solution of the radial Schrödinger equation the approximate formula

$$\psi = z^{m/4} \exp\left(-\frac{c}{m/2 - 1} z^{-(m/2-1)}\right). \tag{3.2.4}$$

3.3 Reflectionless potential

When the energy is given, the potential $V(z) = V_0/\cosh^2(z/a)$, where a is a given constant with the dimension of length, is reflectionless for particular values of V_0. Find such a value of V_0 by introducing into the q-equation (2.1.4) the 'ansatz'

$$q = \frac{k}{1 - 1/[b\cosh^2(z/a)]}, \quad k = (2mE/\hbar^2)^{1/2}, \tag{3.3.1a,b}$$

where b is a dimensionless constant, and determining this constant such that the q-equation is exactly satisfied, which implies that $q^{-1/2}(z)\exp[+i\int^z q(z)dz]$ is an exact solution of the Schrödinger equation, and that hence no reflection occurs.

Solution. The function $R(z)$ in the Schrödinger equation (2.1.1) is

$$R(z) = k^2 - (2mV_0/\hbar^2)/\cosh^2(z/a). \tag{3.3.2}$$

The q-equation (2.1.4) can be written as

$$R(z) = q^2 \left\{ 1 + \left[\frac{d}{dz}\left(\frac{1}{2q}\right)\right]^2 - \frac{1}{q}\frac{d^2}{dz^2}\left(\frac{1}{2q}\right) \right\}. \tag{3.3.3}$$

Inserting (3.3.1a) into (3.3.3), we obtain after straightforward calculations

$$R(z) = \frac{k^2}{\{1 - 1/[b\cosh^2(z/a)]\}^2}$$

$$\times \left[1 + \frac{2}{ba^2k^2\cosh^2(z/a)} - \frac{3b+1}{b^2a^2k^2\cosh^4(z/a)} + \frac{2}{b^2a^2k^2\cosh^6(z/a)}\right]. \tag{3.3.4}$$

Now we write (3.3.4) as

$$R(z) = \frac{k^2}{\{1 - 1/[b\cosh^2(z/a)]\}^2}\left\{\left[1 - \frac{1}{b\cosh^2(z/a)}\right]^2\left[1 + \frac{2}{a^2k^2\cosh^2(z/a)}\right]\right.$$

$$\left. + \frac{2(1 + a^2k^2 - b)}{ba^2k^2\cosh^2(z/a)}\left[1 - \frac{1}{2b\cosh^2(z/a)}\right]\right\}. \tag{3.3.5}$$

If we choose

$$b = 1 + a^2k^2, \tag{3.3.6}$$

(3.3.5) simplifies to

$$R(z) = k^2 + \frac{2}{a^2\cosh^2(z/a)}. \tag{3.3.7}$$

Comparing (3.3.7) and (3.3.2), we obtain

$$V_0 = -\hbar^2/(ma^2). \tag{3.3.8}$$

The q-equation is thus exactly satisfied by (3.3.1a) along with (3.3.6) and (3.3.8), which implies that the potential $-\hbar^2/[ma^2\cosh^2(z/a)]$ is reflectionless.

3.4 Stokes and anti-Stokes lines

Show that:

(a) $dy/dx = -\tan\chi$ on an anti-Stokes line and $dy/dx = \cot\chi$ on a Stokes line, where $z = x + iy$ and $\chi = \arg q(z)$ with $q(z)$ not zero or singular;
(b) there emerge three first-order anti-Stokes lines and three first-order Stokes lines from a simple zero of $Q^2(z)$;
(c) on a curve $z = z(s)$, where s is a real parameter, the absolute value of $\exp[iw(z)]$ assumes its extreme values when $\text{Im}[q(z)dz/ds] = 0$. Use this relation to find, for the first-order approximation when $Q^2(z) = z$, the location of the extreme values of the absolute value of $\exp[iw(z)]$ on a circle with its centre at the zero of $Q^2(z)$.

Solution. (a) According to the definitions (2.3.21a,b)

$$dw = q(z)dz \text{ is real along an anti-Stokes line,} \tag{3.4.1a}$$

$$dw = q(z)dz \text{ is purely imaginary along a Stokes line.} \tag{3.4.1b}$$

Consider a point z where $q(z)$ is not zero or singular and put

$$q(z) = |q(z)|\exp[i\chi(z)], \tag{3.4.2a}$$

$$z = x + iy. \tag{3.4.2b}$$

Then

$$dw = q(z)dz = |q(z)|(\cos\chi + i\sin\chi)(dx + idy)$$
$$= |q(z)|\,[\cos\chi\,dx - \sin\chi\,dy + i(\sin\chi\,dx + \cos\chi\,dy)]. \qquad (3.4.3)$$

Anti-Stokes line: $\operatorname{Im} dw = 0 \Rightarrow dy/dx = -\tan\chi,$
$$dw = |q(z)|(\cos\chi\,dx - \sin\chi\,dy), \qquad (3.4.4)$$
Stokes line: $\operatorname{Re} dw = 0 \Rightarrow dy/dx = \cot\chi, \quad dw = i|q(z)|(\sin\chi\,dx + \cos\chi\,dy).$
$$(3.4.5)$$

From (3.4.4) and (3.4.5) it follows that, at points where $q(z)$ is not zero or singular, the Stokes and anti-Stokes lines are orthogonal to each other.

(b) Without essential loss of generality we can assume that the zero of $Q^2(z)$ lies at the origin and that in the neighbourhood of the origin

$$Q^2(z) = z. \qquad (3.4.6)$$

With $z = r\exp(i\theta)$ we obtain from (3.4.6)

$$Q(z) = z^{1/2} = r^{1/2}\exp(i\theta/2). \qquad (3.4.7)$$

Hence,

$$w(z) = \int_0^z z^{1/2}\,dz = \tfrac{2}{3}z^{3/2} = \tfrac{2}{3}r^{3/2}\exp(i3\theta/2)$$

$$= \tfrac{2}{3}r^{3/2}\cos(3\theta/2) + i\tfrac{2}{3}r^{3/2}\sin(3\theta/2). \qquad (3.4.8)$$

Anti-Stokes lines: $\operatorname{Im} dw = 0 \Rightarrow \operatorname{Im} w = \text{const} \Rightarrow r^{3/2}\sin(3\theta/2) = \text{const}.$
$$(3.4.9)$$

The anti-Stokes lines through the point $r = 0$ are thus determined by the equation

$$r^{3/2}\sin(3\theta/2) = 0, \qquad (3.4.10)$$

from which it follows that $3\theta/2$ is a multiple of π. Hence, three first-order anti-Stokes lines emerge from the zero of $Q^2(z)$ at the angles $\theta = 0, 2\pi/3$ and $4\pi/3$.

Stokes lines: $\operatorname{Re} dw = 0 \Rightarrow \operatorname{Re} w = \text{const} \Rightarrow r^{3/2}\cos(3\theta/2) = \text{const}.$
$$(3.4.11)$$

The Stokes lines through the point $r = 0$ are thus determined by the equation

$$r^{3/2}\cos(3\theta/2) = 0, \qquad (3.4.12)$$

from which it follows that $3\theta/2 - \pi/2$ is a multiple of π, i.e., that first-order Stokes lines emerge from the zero of $Q^2(z)$ at the angles $\theta = \pi/3, \pi$ and $5\pi/3$. Hence, three first-order Stokes lines emerge from the zero of $Q^2(z)$, bisecting the angles between the three emerging anti-Stokes lines.

If, more generally, $Q^2(z) = Cz$, where C is a complex constant, the only change is a rotation of the anti-Stokes and the Stokes lines by the angle $-(1/3) \arg C$, i.e., a rotation in the negative or the positive sense, depending on whether $\arg C$ is positive or negative.

(c) Consider the absolute value of $\exp[iw(z)]$ on the curve $z = z(s)$, where s is a real parameter. Since $dw/dz = q(z)$, we obtain

$$\frac{d}{ds}|\exp[iw(z)]| = \frac{d}{ds}\exp[-\mathrm{Im}\, w(z)] = -\exp[-\mathrm{Im}\, w(z)]\frac{d}{ds}\mathrm{Im}\, w(z)$$

$$= -\exp[-\mathrm{Im}\, w(z)]\,\mathrm{Im}\left(\frac{dw(z)}{dz}\frac{dz}{ds}\right)$$

$$= -|\exp[iw(z)]|\,\mathrm{Im}[q(z)dz/ds]. \qquad (3.4.13)$$

Thus, $|\exp[iw(z)]|$ assumes its extreme values when

$$\mathrm{Im}[q(z)dz/ds] = 0. \qquad (3.4.14)$$

Special case. Let $Q^2(z) = z$ and let the curve $z = z(s)$ be a circle with its centre at the origin; see Fig. 3.4.1. Thus $z = z(\theta) = r\exp(i\theta)$, where r is a constant, and the parameter s is the angle θ. Since $dz/d\theta = ir\exp(i\theta)$, condition (3.4.14) with $q(z) = Q(z)$ and $s = \theta$ gives

$$\mathrm{Im}[i\,Q(z)\exp(i\theta)] = 0. \qquad (3.4.15)$$

Whether a certain extreme value corresponds to a maximum or a minimum of the

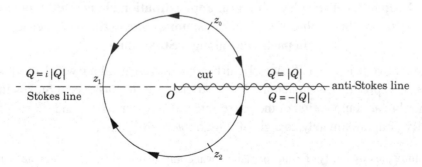

Figure 3.4.1 Arrows indicate the directions in which the absolute value of $\exp[iw(z)]$ increases, when the first-order approximation is used, and $Q(z)$ is given by (3.4.7). Thus there are minima at z_0 and z_2 and a maximum at z_1.

absolute value of $\exp[i\,w(z)]$ depends on the choice of phase of $Q(z)$. If $Q(z)$ is given by (3.4.7), condition (3.4.15) takes the form

$$\mathrm{Im}[i\exp(i3\theta/2)] = 0 \quad \Rightarrow \quad \cos(3\theta/2) = 0 \quad \Rightarrow \quad \theta = \pi/3,\ \pi,\ 5\pi/3. \quad (3.4.16)$$

3.5 Properties of the phase-integral approximation along an anti-Stokes line

Consider the situation in which the points z_0 and z and the lower limit of integration in the integral (2.1.3b) defining $w(z)$ lie on the same anti-Stokes line. A certain linear combination of the phase-integral functions $f_1(z)$ and $f_2(z)$ is fitted, exactly or approximately, to a given solution ψ at the point z_0. What are the restrictions for representing this solution by the same linear combination of phase-integral functions when one moves from z_0 to z?

Solution. The given conditions imply that $|\exp[i\,w(z)]| = |\exp[i\,w(z_0)]| = 1$, and from the basic estimates (2.3.17a–d) we conclude that, if $\mu(z, z_0)$, defined in (2.3.18), is much smaller than unity, the matrix $\mathbf{F}(z, z_0)$ is approximately a unit matrix, and the a-coefficients in a linear combination of the phase-integral functions are approximately constant. When we have fitted, exactly or approximately, a linear combination of the phase-integral functions to a given wave function $\psi(z)$ at the point z_0 on the anti-Stokes line, the same linear combination is a good approximation for $\psi(z)$ along the whole anti-Stokes line, in both directions from z_0, as long as we do not approach transition points, i.e., zeros or singularities of $Q^2(z)$, causing the condition $\mu(z, z_0) \ll 1$ to be violated. In particular, if the potential is real on the real z-axis, a classically allowed region on the real z-axis is an anti-Stokes line, and the phase-integral approximation there has the simple properties just described.

3.6 Properties of the phase-integral approximation along a path on which the absolute value of $\exp[i\,w(z)]$ is monotonic in the strict sense, in particular along a Stokes line

Consider two points z_0 and z, which can be joined by a path along which the absolute value of $\exp[i\,w(z)]$ is monotonic in the strict sense, for instance a Stokes line. A certain linear combination of the phase-integral functions $f_1(z)$ and $f_2(z)$ is fitted, exactly or approximately, to a given solution ψ at the point z_0.

(a) Study the possibility of using the same linear combination to represent ψ, when z moves away from z_0 in the direction of:
  (i) increasing absolute value of $\exp[i\,w(z)]$;
  (ii) decreasing absolute value of $\exp[i\,w(z)]$.
(b) Formulate a rule of thumb for using the phase-integral approximation.

Solution. At the point z_0 the solution $\psi(z)$ is given by (2.3.4a) along with (2.3.1a,b), i.e.,

$$\psi(z_0) = a_1(z_0)q^{-1/2}(z_0)\exp[i\,w(z_0)] + a_2(z_0)q^{-1/2}(z_0)\exp[-i\,w(z_0)]. \quad (3.6.1)$$

When z does not move along an anti-Stokes line, the properties of the phase-integral approximation are complicated, and caution must be exercised even in an interval where the μ-integral (2.3.18) is small compared with unity. If z moves away from the point z_0 on a curve in the complex z-plane along which the absolute value of $\exp[i\,w(z)]$ is increasing in the strict sense, we realize from the basic estimates (2.3.17a–d) that, since the absolute value of $\exp[i\,w(z)]$ soon becomes large compared with unity, the condition $\mu(z, z_0) \ll 1$ no longer ensures that the matrix $\mathbf{F}(z, z_0)$ is approximately a unit matrix. This is also true when $\exp[i\,w(z)]$ decreases in the direction from z_0 to z, since then the absolute value of $\exp[-i\,w(z)]$ soon becomes large and the coefficients $a_1(z)$ and $a_2(z)$ interchange their properties.

(a)(i) Letting z move away from z_0 along the curve of monotonicity in the direction in which the absolute value of $\exp[+i\,w(z)]$ increases, we obtain from (2.3.7)

$$|a_1(z) - a_1(z_0)| \le |F_{11}(z, z_0) - 1||a_1(z_0)| + |F_{12}(z, z_0)a_2(z_0)|, \quad (3.6.2a)$$
$$|a_2(z) - a_2(z_0)| \le |F_{21}(z, z_0)a_1(z_0)| + |F_{22}(z, z_0) - 1||a_2(z_0)|. \quad (3.6.2b)$$

From these inequalities and the basic estimates (2.3.17a–d) we can draw the following conclusions as to the behaviour of $a_1(z)$ and $a_2(z)$.

If

$$|a_2(z_0)\exp[-i\,w(z_0)]| \lesssim |a_1(z_0)\exp[i\,w(z_0)]|, \quad (3.6.3)$$

i.e., if the order of magnitude of $|a_2(z_0)f_2(z_0)|$ is less than or at the most the same as that of $|a_1(z_0)f_1(z_0)|$, it follows from (3.6.2a), (3.6.3) and (2.3.17a,b) that

$$|a_1(z) - a_1(z_0)| = O(\mu)|a_1(z_0)|, \quad (3.6.4)$$

and $a_1(z)$ is thus approximately constant. However, the coefficient $a_2(z)$ differs considerably from $a_2(z_0)$ unless z lies very close to z_0, $|a_2(z)/a_2(z_0)|$ being in general much larger than unity. The term $a_2(z)f_2(z)$ is, however, insignificant compared with $a_1(z)f_1(z)$ when $a_1(z_0) \ne 0$, unless z lies very close to z_0; see (3.6.6). Thus, the fact that $a_2(z)$ changes greatly when the point z moves away from z_0 does not spoil the possibility of using the phase-integral approximation to represent $\psi(z)$ along a path from z_0 towards increasing absolute value of $\exp[+i\,w(z)]$, as long as $\mu(z, z_0) \ll 1$. In fact, according to (2.3.4a), (2.3.1a,b) and (2.3.7) we obtain

$$\psi(z) = q^{-1/2}(z)\exp[i\,w(z)](\{F_{11}(z, z_0) + F_{21}(z, z_0)\exp[-2i\,w(z)]\}a_1(z_0)$$
$$+ \{F_{12}(z, z_0) + F_{22}(z, z_0)\exp[-2i\,w(z)]\}a_2(z_0)), \quad (3.6.5)$$

and hence the basic estimates (2.3.17a–d) and the condition (3.6.3) yield

$$\frac{\psi(z)}{a_1(z_0)q^{-1/2}(z)\exp[i\,w(z)]} - 1$$

$$= O(\mu) + \{O(\mu)\exp[-2i\,w(z_0)] + O(1)\exp[-2i\,w(z)]\}\frac{a_2(z_0)}{a_1(z_0)}$$

$$= O(\mu) + O(1)\exp\{-2i[w(z) - w(z_0)]\}\,. \tag{3.6.6}$$

The wave function is thus given approximately by

$$\psi(z) = a_1(z_0)q^{-1/2}(z)\exp[i\,w(z)] = a_1(z_0)f_1(z) \tag{3.6.7}$$

with a relative error at the most of the order of magnitude $\mu(z, z_0)$, if condition (3.6.3) is fulfilled, and if, furthermore, the absolute value of $\exp\{-2i[w(z) - w(z_0)]\}$ is less than or at the most of the order of magnitude μ, which means that z must not lie too close to z_0. If z lies close to z_0, the term $a_2(z_0)f_2(z)$ is significant and must be retained, but sufficiently far away from z_0 it is insignificant.

On the other hand, if the condition (3.6.3) is *not* fulfilled, i.e., if

$$|a_2(z_0)\exp[-i\,w(z_0)]| \gg |a_1(z_0)\exp[i\,w(z_0)]|, \tag{3.6.8}$$

we cannot use the phase-integral approximation in the direction in which the absolute value of $\exp[+i\,w(z)]$ increases, unless z lies very close to z_0.

(a)(ii) If z moves on a path along which the absolute value of $\exp[+i\,w(z)]$ decreases monotonically, which means that the absolute value of $\exp[-i\,w(z)]$ increases monotonically, the coefficients $a_1(z)$ and $a_2(z)$ interchange their properties as compared with the case treated in (a)(i). We can then use the phase-integral approximation to represent $\psi(z)$ away from z_0 only if

$$|a_1(z_0)\exp[i\,w(z_0)]| \lesssim |a_2(z_0)\exp[-i\,w(z_0)]|. \tag{3.6.9}$$

If this condition is fulfilled, and z does not lie close to z_0, we have

$$\psi(z) \approx a_2(z_0)q^{-1/2}(z)\exp[-i\,w(z)] = a_2(z_0)f_2(z), \tag{3.6.10}$$

but if z lies sufficiently close to z_0, the term $a_1(z_0)f_1(z)$ is significant and must be retained.

(b) Our results as to the properties and validity of the phase-integral approximation along a path on which the absolute value of $\exp[i\,w(z)]$ increases or decreases rapidly can be summarized as a rule of thumb: If at the point z_0 the wave function is given by (3.6.1), and if both terms in (3.6.1) are of the same order of magnitude at $z = z_0$, then both of the conditions (3.6.3) and (3.6.9) are fulfilled, and the phase-integral approximation can be used to represent $\psi(z)$ in both directions from z_0, as long as $\mu(z, z_0) \ll 1$. As one moves away from the point z_0, keeping $a_1(z_0)$ and $a_2(z_0)$ constant, and, of course, replacing z_0 at the other places in (3.6.1) by the moving coordinate z, both terms in (3.6.1) are at first significant, but very soon the term containing the decreasing exponential becomes insignificant, and then $\psi(z)$ is given by (3.6.7) in the direction of increasing absolute value of $\exp[+iw(z)]$ and by (3.6.10) in the direction of increasing absolute value of $\exp[-iw(z)]$. If, however, the absolute value of one of the terms in (3.6.1) is much larger than the absolute value of the other term, then only one of the conditions (3.6.3) and (3.6.9) is fulfilled, and we can use the phase-integral approximation to represent $\psi(z)$ away fom z_0 only in the direction in which the exponential factor of the term that dominates at the point z_0 increases. We may say in short, albeit less precisely, that on a Stokes line one must not proceed with the phase-integral approximation in the direction of decreasing absolute value of $\psi(z)$.

Already the properties of the phase-integral approximation along a Stokes line, analysed as a particular case in this problem, imply that the connection formulas associated with a turning point, to be derived in Section 3.10 and Section 3.12, are one-directional. The discussion in the present problem is also important for other problems; in particular it forms the basis for the treatment in Section 3.28.

3.7 Determination of the Stokes constants associated with the three anti-Stokes lines that emerge from a well-isolated, simple transition zero

When z moves in the complex z-plane a full turn along a closed curve that does not enclose any singularity of $R(z)$, the exact wave function remains single-valued. Use this fact to determine the Stokes constants associated with the three anti-Stokes lines that emerge from a simple transition zero, i.e., a first-order zero of $Q^2(z)$, lying so far away from all other transition points that there exists a contour enclosing the zero in question, but no other transition point or singularity of $R(z)$, along which the μ-integral is small compared with unity.

Solution. We define $w(z)$ according to (2.4.1), i.e.,

$$w(z) = \int\limits_{(t)}^{z} q(z)dz, \tag{3.7.1}$$

where t is the transition zero. When $w(z)$ is defined in this way, the single-valuedness of the wave function as z moves a full turn around t implies according to (2.3.16) that

$$\mathbf{F}(z_3, z_0) = \begin{pmatrix} 0 & i \\ i & 0 \end{pmatrix}, \tag{3.7.2}$$

where z_0 and z_3 are points shown in Fig. 3.7.1. On a path, joining the points z_v and z_{v+1} (where v may be equal to 0, 1 or 2) in Fig. 3.7.1(a) and (b), the absolute value

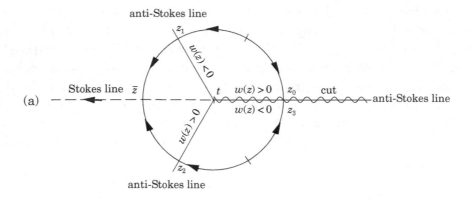

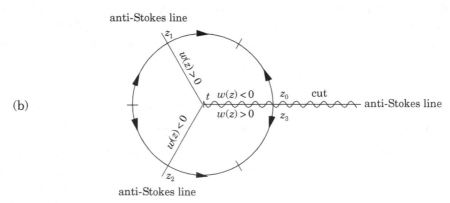

Figure 3.7.1 Anti-Stokes lines (full) emerging from a simple zero t of $Q^2(z)$. The difference between (a) and (b) is due to different choices of the phase of $Q(z)$, as is seen from the value of $w(z)$ defined according to (2.4.1), on the anti-Stokes lines. Arrows indicate directions in which the absolute value of $\exp[i w(z)]$ increases. The dashed line $t\bar{z}$ in (a) is a Stokes line that will be used in Sections 3.8 and 3.9. The figure is drawn for the case of first-order anti-Stokes lines emerging from t. When one uses a higher order of the phase-integral approximation, one *defines* anti-Stokes lines emerging from t as explained at the end of Subsection 2.3.5.

of $\exp[i w(z)]$ has either a single minimum or a single maximum, and the absolute values of $\exp[\pm i w(z_v]$ and $\exp[\pm i w(z_{v+1})]$ are equal to unity. Hence the formulas (2.3.20a,b) apply.

For the situation in Fig. 3.7.1(a) there is a minimum on the paths $z_0 z_1$ and $z_2 z_3$, but a maximum on the path $z_1 z_2$. Thus, according to (2.3.20a,b) we have

$$\mathbf{F}(z_3, z_2) \approx \begin{pmatrix} 1 & a \\ 0 & 1 \end{pmatrix}, \tag{3.7.3a}$$

$$\mathbf{F}(z_2, z_1) \approx \begin{pmatrix} 1 & 0 \\ b & 1 \end{pmatrix}, \tag{3.7.3b}$$

$$\mathbf{F}(z_1, z_0) \approx \begin{pmatrix} 1 & c \\ 0 & 1 \end{pmatrix}, \tag{3.7.3c}$$

where a, b and c are the Stokes constants that are to be determined. Using the multiplication rule (2.3.12b), we obtain with the aid of (3.7.3a–c)

$$\mathbf{F}(z_3, z_0) = \mathbf{F}(z_3, z_2)\mathbf{F}(z_2, z_1)\mathbf{F}(z_1, z_0) \approx \begin{pmatrix} 1+ab & a+c+abc \\ b & 1+bc \end{pmatrix}. \tag{3.7.4}$$

Identifying the two expressions (3.7.2) and (3.7.4) for $\mathbf{F}(z_3, z_0)$, we get the equations

$$1 + ab = 0, \tag{3.7.5a}$$
$$a + c + abc = i, \tag{3.7.5b}$$
$$b = i, \tag{3.7.5c}$$
$$1 + bc = 0, \tag{3.7.5d}$$

which give

$$a = b = c = i. \tag{3.7.6}$$

For the situation in Fig. 3.7.1(b) there is a maximum on the paths $z_0 z_1$ and $z_2 z_3$, but a minimum on the path $z_1 z_2$. Thus, according (2.3.20a,b) we have

$$\mathbf{F}(z_3, z_2) \approx \begin{pmatrix} 1 & 0 \\ a' & 1 \end{pmatrix}, \tag{3.7.7a}$$

$$\mathbf{F}(z_2, z_1) \approx \begin{pmatrix} 1 & b' \\ 0 & 1 \end{pmatrix}, \tag{3.7.7b}$$

$$\mathbf{F}(z_1, z_0) \approx \begin{pmatrix} 1 & 0 \\ c' & 1 \end{pmatrix}, \tag{3.7.7c}$$

where a', b' and c' are the Stokes constants that are to be determined. Calculations analogous to those made above for the situation in Fig. 3.7.1(a) yield

$$a' = b' = c' = i. \tag{3.7.8}$$

The Stokes constant, F_{12} or F_{21}, associated with two neighbouring anti-Stokes lines, emerging from a simple transition zero lying far away from all other transition points, is thus equal to i.

3.8 Connection formula for tracing a phase-integral wave function from a Stokes line emerging from a simple transition zero t to the anti-Stokes line emerging from t in the opposite direction

Use the results obtained in Section 3.7 to derive the connection formula in question.

Solution. Suppose that at the point $\bar{z}$ on the Stokes line in Fig. 3.7.1(a) a solution ψ is represented by a phase-integral expression with $a_1(\bar{z}) = 0$ and $a_2(\bar{z}) = 1$, i. e.,

$$\psi(\bar{z}) = f_2(\bar{z}) = q^{-1/2}(\bar{z})\exp[-iw(\bar{z})], \tag{3.8.1a}$$

$$w(\bar{z}) = \int\limits_{(t)}^{\bar{z}} q(z)dz. \tag{3.8.1b}$$

This solution remains approximately valid in the direction of increasing absolute value of $\exp[-iw(z)]$. Along the Stokes line this is in the direction towards the transition zero t, i.e., the zero of $Q^2(z)$, and along the circular bow up to the point z_1, which lies on an anti-Stokes line emerging from t. Hence we have approximately $\psi(z_1) = f_2(z_1)$, and we can thus put

$$a_1(z_1) = 0, \tag{3.8.2a}$$

$$a_2(z_1) = 1. \tag{3.8.2b}$$

Inverting the matrix given by (3.7.3c) and (3.7.6), we obtain

$$\mathbf{F}(z_0, z_1) = \begin{pmatrix} 1 & -i \\ 0 & 1 \end{pmatrix}. \tag{3.8.3}$$

With the aid of (3.8.2a,b) and (3.8.3) we obtain according to (2.3.7)

$$\begin{pmatrix} a_1(z_0) \\ a_2(z_0) \end{pmatrix} = \mathbf{F}(z_0, z_1) \begin{pmatrix} a_1(z_1) \\ a_2(z_1) \end{pmatrix} = \begin{pmatrix} -i \\ 1 \end{pmatrix} \tag{3.8.4}$$

and hence

$$\psi(z_0) = -iq^{-1/2}(z_0)\exp[+iw(z_0)] + q^{-1/2}(z_0)\exp[-iw(z_0)]$$
$$= 2\exp(-i\pi/4)q^{-1/2}(z_0)\cos[w(z_0) - \pi/4]. \tag{3.8.5}$$

We can write (3.8.1a) and (3.8.5) as the following one-directional connection formula for the connection from the Stokes line to the upper lip of the cut along the anti-Stokes line in Fig. 3.7.1(a):

$$q^{-1/2}(z)\exp[-iw(z)] \rightarrow 2\exp(-i\pi/4)q^{-1/2}(z)\cos[w(z) - \pi/4], \tag{3.8.6}$$

where the absolute value of $\exp[-iw(z)]$ increases along the Stokes line in the direction towards the transition zero t, $w(z)$ is given by (3.8.1b) and the arrow indicates the direction in which the inference can be drawn.

In particular the transition zero can be a turning point on the real z-axis, in which case the Stokes and anti-Stokes lines under consideration coincide with parts of the real z-axis. As a particular case of (3.8.6) we then obtain the connection formula (3.10.10′) in Section 3.10.

3.9 Connection formula for tracing a phase-integral wave function from an anti-Stokes line emerging from a simple transition zero t to the Stokes line emerging from t in the opposite direction

Use the results obtained in Section 3.7 to derive the connection formula in question.

Solution. Assume that the solution ψ at the point z_0 on the upper lip of the cut along the anti-Stokes line in Fig. 3.7.1(a) is given by

$$\psi(z_0) = a_1(z_0)f_1(z_0) + a_2(z_0)f_2(z_0), \tag{3.9.1}$$

where the phase-integral functions are given by (2.3.1a,b) with $w(z)$ defined by (3.7.1). According to (3.7.3c) and (3.7.6) we have

$$\mathbf{F}(z_1, z_0) = \begin{pmatrix} 1 & i \\ 0 & 1 \end{pmatrix}. \tag{3.9.2}$$

Using this matrix in the general formula expressing the a-coefficients at z_1 in terms of those at z_0, we obtain

$$\begin{pmatrix} a_1(z_1) \\ a_2(z_1) \end{pmatrix} = \mathbf{F}(z_1, z_0) \begin{pmatrix} a_1(z_0) \\ a_2(z_0) \end{pmatrix} = \begin{pmatrix} a_1(z_0) + ia_2(z_0) \\ a_2(z_0) \end{pmatrix}, \tag{3.9.3}$$

and hence from (2.3.4a)

$$\psi(z_1) = a_1(z_1)f_1(z_1) + a_2(z_1)f_2(z_1) = [a_1(z_0) + ia_2(z_0)]f_1(z_1) + a_2(z_0)f_2(z_1). \tag{3.9.4}$$

On tracing this solution from z_1 to $\bar{z}$ in Fig. 3.7.1(a) by noting that the absolute value of $\exp[iw(z)]$ is monotonically increasing from z_1 to $\bar{z}$, one finds that the term in (3.9.4) containing $f_2(z_1)$ is insignificant. Thus, one has approximately

$$\psi(\bar{z}) = [a_1(z_0) + ia_2(z_0)]f_1(\bar{z}). \tag{3.9.5}$$

Recalling (2.3.1a,b) and putting $a_1(z_0) = A$ and $a_2(z_0) = B$, we can write (3.9.1) and (3.9.5) as the following connection formula

$$Aq^{-1/2}(z)\exp[+iw(z)] + Bq^{-1/2}(z)\exp[-iw(z)]$$
$$\to (A + iB)q^{-1/2}(z)\exp[+iw(z)], \tag{3.9.6}$$

for tracing a solution from the upper lip of the cut along the anti-Stokes line in Fig. 3.7.1(a) to the opposite Stokes line. This connection formula is valid provided that $A + iB$ is not too close to zero, and that the phase of $q(z)$ is chosen such that the absolute value of $\exp[iw(z)]$ has the monotonicity properties shown by the arrows in Fig. 3.7.1(a). The properties of the phase-integral approximation along a Stokes line obviously imply that the connection formula (3.9.6) is one-directional; see Section 3.6.

If in particular $R(z)$ and $Q^2(z)$ are real on the real z-axis, and the transition zero is a turning point on this axis, in which case the Stokes and anti-Stokes lines under consideration coincide with parts of the real z-axis, we obtain as a particular case of (3.9.6) the connection formula (3.12.6).

3.10 Connection formula for tracing a phase-integral wave function from a classically forbidden to a classically allowed region

This connection formula, which applies when the coefficient function $R(z)$ and $Q^2(z)$ are real on the real z-axis, is readily obtained as a special case of the connection formula (3.8.6) pertaining to the more general situation that $R(z)$ and $Q^2(z)$ need not be real on the real axis and the transition zero may lie in the complex z-plane.

Use the exact parameterized F-matrix presented in Section 2.4 for an alternative derivation that displays explicitly the conditions for the usefulness of the formula.

Solution. We assume the wave function to be real on the real z-axis, which is not an essential restriction, and, as in Section 2.4, we denote a point in the classically forbidden region by x_1 and a point in the classically allowed region by x_2, both these points being assumed to lie sufficiently far away from the turning point t in order that the phase-integral approximation be valid. The phase integral is defined according to (2.4.1), i.e.,

$$w(z) = \int_{(t)}^{z} q(z)dz. \tag{3.10.1}$$

If the wave function at the point x_1 in the classically forbidden region is given by (2.4.23), i.e.,

$$\psi(x_1) = C(x_1)\left|q^{-1/2}(x_1)\right|\exp[+|w(x_1)|] + D(x_1)\left|q^{-1/2}(x_1)\right|\exp[-|w(x_1)|], \tag{3.10.2}$$

with $C(x_1)$ and $D(x_1)$ real, the wave function at the point x_2 in the classically allowed region is obtained from (2.4.26), i.e.,

$$\psi(x_2) = A(x_2)\left|q^{-1/2}(x_2)\right|\exp(+i|w(x_2)|) + B(x_2)\left|q^{-1/2}(x_2)\right|\exp(-i|w(x_2)|), \tag{3.10.3}$$

where according to (2.4.28a,b)

$$A(x_2) = \left\{ \frac{1}{2\alpha} C(x_1) + i\alpha[\gamma C(x_1) - D(x_1)] \right\} \exp[+i(\pi/4 - \beta)], \qquad (3.10.4a)$$

$$B(x_2) = A^*(x_2). \qquad (3.10.4b)$$

With α, β and γ defined in terms of F-matrix elements according to (2.4.7a–c) the formulas displayed above are exact. In order to obtain useful approximate formulas we recall that the quantities α, β and γ are real and fulfil the estimates (2.4.8a–c), i.e.,

$$\alpha = \alpha(x_1, x_2) = 1 + O(\mu), \qquad (3.10.5a)$$

$$\beta = \beta(x_1, x_2) = O(\mu), \qquad (3.10.5b)$$

$$\gamma = \gamma(x_1, x_2) = O(\mu)\exp(2|w(x_1)|), \qquad (3.10.5c)$$

where μ is assumed to be small compared with unity. In general μ is large compared with $\exp(-2|w(x_1)|)$, and therefore the exponential function in (3.10.5c) is in general so large that $|\gamma| \gg 1$; see (2.4.6).

Since γ is unknown and in general large compared with unity, it follows from (3.10.4a,b) and (3.10.5a,b) that we can obtain approximate expressions for $A(x_2)$ and $B(x_2)$ only when

$$|\gamma C(x_1)| \ll |D(x_1)|. \qquad (3.10.6)$$

According to (3.10.5c) this requirement is fulfilled if $\mu \ll 1$ and

$$|C(x_1)/D(x_1)| \lesssim \exp[-2|w(x_1)|], \qquad (3.10.7)$$

which means that at the point x_1 the term in the phase-integral expression for the wave function that increases in the direction towards the turning point is at least of the same order of magnitude as the decreasing term. When $\exp[-2|w(x_1)|] \ll 1$ we then obtain from (3.10.4a,b), (3.10.5a,b), (3.10.6) and (3.10.7) the approximate formulas

$$A(x_2) = D(x_1)\exp(-i\pi/4), \qquad (3.10.8a)$$

$$B(x_2) = D(x_1)\exp(+i\pi/4). \qquad (3.10.8b)$$

Inserting (3.10.8a,b) into (3.10.3), we obtain

$$\psi(x_2) = 2D(x_1)\left|q^{-1/2}(x_2)\right| \cos[|w(x_2)| - \pi/4]. \qquad (3.10.9)$$

Now we write (3.10.2) and (3.10.9) in the form of the approximate, but in general very accurate, connection formula

$$C\left|q^{-1/2}(x)\right|\exp[+|w(x)|] + D\left|q^{-1/2}(x)\right|\exp[-|w(x)|]$$

$$\to 2D\left|q^{-1/2}(x)\right| \cos[|w(x)| - \pi/4], \qquad (3.10.10)$$

which is valid under condition (3.10.7), which means that on the left-hand side of (3.10.10) the order of magnitude of the first term must not exceed that of the second term. In the literature one usually finds the particular case of (3.10.10) that corresponds to $C = 0$ and $D = 1$, i.e.,

$$|q^{-1/2}(x)|\exp[-|w(x)|] \to 2|q^{-1/2}(x)|\cos[|w(x)| - \pi/4]. \quad (3.10.10')$$

This connection formula, which is a particular case of (3.8.6), was first obtained by Jeffreys (1925) for the first-order WKB approximation and by N. Fröman (1970) for an arbitrary order of the phase-integral approximation under the assumption that $Q^2(z) = R(z)$, but according to the treatment in this section this assumption is not necessary. It is seen that the connection formula has the same form for every order of the phase-integral approximation, but it was not clear that this should be the case until the connection formula had actually been derived for an arbitrary order of approximation. The one-directional nature of the connection formula (3.10.10') has been discussed in detail and even confirmed numerically by N. Fröman (1966a); see fig. 6 in that paper.

If, in particular, the wave function is subject to the condition of being equal to zero at the point x_1 in the classically forbidden region, one obtains from (3.10.2) $C(x_1)/D(x_1) = -\exp[-2|w(x_1)|]$. Condition (3.10.7) is thus fulfilled, and hence the wave function at a point x_2 in the classically allowed region is approximately given by (3.10.9) and is approximately independent of the location of the point x_1, where it is required to be equal to zero. See, however, Section 3.24.

It is instructive also to consider the wave function in the classically allowed region when condition (3.10.6) for the validity of the connection formula (3.10.10) is violated, such that instead the condition

$$|\gamma C(x_1)| \gg |D(x_1)| \quad (3.10.11)$$

prevails. We then obtain from (3.10.4a,b) and (3.10.5a,b) the approximate formulas

$$A(x_2) = \left[\tfrac{1}{2}\exp(+i\pi/4) + \gamma\exp(+3i\pi/4)\right]C(x_1), \quad (3.10.12a)$$

$$B(x_2) = \left[\tfrac{1}{2}\exp(-i\pi/4) + \gamma\exp(-3i\pi/4)\right]C(x_1), \quad (3.10.12b)$$

with the aid of which we obtain from (3.10.3), when $|\gamma| \gg 1/2$, the approximate formula

$$\psi(x_2) = 2C(x_1)\gamma|q^{-1/2}(x_2)|\cos[|w(x_2)| + 3\pi/4]$$

$$= 2C(x_1)|\gamma q^{-1/2}(x_2)|\cos[|w(x_2)| + (1 + 2\gamma/|\gamma|)\pi/4], \quad |\gamma| \gg 1/2. \quad (3.10.13)$$

Since in (3.10.13) the absolute value of γ is large compared with unity and unknown (even as concerns its sign), the wave function in the classically allowed region has a large and unknown amplitude, but a phase that is approximately known modulo π. The reader who is interested to see how rapidly the phase and amplitude of the wave function change when γ changes sign is referred to eqs. (48a,b) and figs. 8a and 8b in N. Fröman (1966a) which illustrate in a concrete way a particular case of (3.10.2) and (3.10.13).

3.11 One-directional nature of the connection formula for tracing a phase-integral wave function from a classically forbidden to a classically allowed region

Demonstrate by the use of the parameterized formulas for the F-matrix presented in Section 2.4, and also by the use of more qualitative arguments, the one-directional nature of the connection formula (3.10.10′) derived in Section 3.10.

Solution. If the wave function at the point x_2 in the classically allowed region is given by the right-hand side of the connection formula (3.10.10′), i.e., by

$$\psi(x_2) = 2\left|q^{-1/2}(x_2)\right| \cos[|w(x_2)| - \pi/4]$$
$$= A(x_2)\left|q^{-1/2}(x_2)\right|\exp[+i|w(x_2)|] + B(x_2)\left|q^{-1/2}(x_2)\right|\exp[-i|w(x_2)|]$$

$$(3.11.1)$$

with

$$A(x_2) = \exp(-i\pi/4), \qquad (3.11.2a)$$
$$B(x_2) = \exp(+i\pi/4), \qquad (3.11.2b)$$

the wave function at the point x_1 in the classically forbidden region is, according to (2.4.23) and (2.4.29a,b),

$$\psi(x_1) = C(x_1)\left|q^{-1/2}(x_1)\right|\exp[+|w(x_1)|] + D(x_1)\left|q^{-1/2}(x_1)\right|\exp[-|w(x_1)|]$$

$$(3.11.3)$$

with

$$C(x_1) = \alpha \exp[-i(\pi/4 - \beta)]A(x_2) + \alpha \exp[+i(\pi/4 - \beta)]B(x_2), \qquad (3.11.4a)$$

$$D(x_1) = \left(\alpha\gamma + \frac{i}{2\alpha}\right)\exp[-i(\pi/4 - \beta)]A(x_2)$$
$$+ \left(\alpha\gamma - \frac{i}{2\alpha}\right)\exp[+i(\pi/4 - \beta)]B(x_2). \qquad (3.11.4b)$$

Inserting (3.11.2a,b) into (3.11.4a,b), we get

$$C(x_1) = 2\alpha \sin \beta, \tag{3.11.5a}$$

$$D(x_1) = \frac{\cos \beta}{\alpha} + 2\alpha\gamma \sin \beta, \tag{3.11.5b}$$

and inserting (3.11.5a,b) into (3.11.3), we get

$$\psi(x_1) = \left|q^{-1/2}(x_1)\right| \exp[-|w(x_1)|]$$
$$\times \left\{ 2\alpha \sin \beta \exp[2|w(x_1)|] + \frac{\cos \beta}{\alpha} + 2\alpha\gamma \sin \beta \right\}. \tag{3.11.6}$$

Using the estimates (2.4.8a–c) for α, β and γ we *cannot* infer that the expression within the braces in (3.11.6) is approximately equal to unity, and it is thus seen that the connection formula (3.10.10′) is one-directional.

We shall now illustrate the one-directional nature of the connection formula (3.10.10′), when the differential equation is the Airy differential equation, i.e.,

$$d^2\psi/dz^2 - z\psi = 0. \tag{3.11.7}$$

The asymptotic formulas for the Airy function Ai(z) are

$$2\pi^{1/2} \, \mathrm{Ai}(z)_{z \to -\infty} 2\left|z^{-1/4}\right| \cos[|\zeta| - \pi/4 - 5/(72|\zeta|)], \tag{3.11.8a}$$

$$2\pi^{1/2} \, \mathrm{Ai}(z)_{z \to +\infty} \left|z^{-1/4}\right| \exp(-|\zeta|), \tag{3.11.8b}$$

where

$$\zeta = \int_0^z z^{1/2} dz = \tfrac{2}{3} z^{3/2}. \tag{3.11.9}$$

When the wave function is given by (3.11.8b) at $z = +\infty$, the wave function is given by (3.11.8a) sufficiently far to the left of the turning point, and when we neglect the phase $-5/(72|\zeta|)$ in (3.11.8a), we obtain the connection formula (3.10.10′) for the first-order approximation. On the other hand, if, at a point z that lies in the classically allowed region and well away from the turning point, we let the wave function be given by $2|z^{-1/4}| \cos(|\zeta| - \pi/4)$, i.e., without the exact expression for the phase, the same wave function tends to infinity as $z \to +\infty$ in the classically forbidden region. For the first-order solutions of the differential equation (3.11.7) this shows that the connection formula (3.10.10′) is one-directional.

One can easily understand the one-directional nature of the connection formula (3.10.10′) by simple examination of the wave function in the classically forbidden region. Consider for this purpose the differential equation

$$d^2\psi/dx^2 - \psi = 0, \tag{3.11.10}$$

the general solution of which is

$$\psi = C \exp(x) + D \exp(-x). \tag{3.11.11}$$

If $C = 0$ (exactly) and $D \neq 0$, the solution is $\psi = D \exp(-x)$, but if C is known to be only approximately equal to zero, the solution for sufficiently large positive values of x will become approximately $\psi = C \exp(x)$, since the term $D \exp(-x)$ soon becomes insignificant. If instead $C \neq 0$ and $D = 0$ (exactly), the solution is $\psi = C \exp(x)$, but if D is known to be only approximately equal to zero, the solution for sufficiently large negative values of x will become approximately $\psi = D \exp(-x)$, since in this case the term $C \exp(x)$ soon becomes insignificant. This simple argumentation demonstrates in a concrete way that the behaviour of the solution in the classically forbidden region implies the one-directional nature of the connection formula (3.10.10′). The one-directional nature of this connection formula is thus a consequence of the simple fact, explained in Section 3.6, that one cannot proceed with an approximate solution, or an exact solution with approximately known initial conditions, in a classically forbidden region from the initial point in the direction in which the wave function decreases. In the context of the numerical solution of a second-order differential equation this shows up in the fact that the solution becomes unstable when one integrates the differential equation in such a direction.

3.12 Connection formulas for tracing a phase-integral wave function from a classically allowed to a classically forbidden region

Use the exact parameterized formulas for the F-matrix presented in Section 2.4 to derive the connection formulas in question, which apply when the coefficient function $R(z)$ and $Q^2(z)$ are real on the real z-axis.

Solution. As in Sections 3.7–3.11 we define the phase-integral according to (2.4.1). We assume the wave function at the point x_2 in the classically allowed region to be given by (2.4.26), i.e.,

$$\psi(x_2) = A\left|q^{-1/2}(x_2)\right|\exp[+i|w(x_2)|] + B\left|q^{-1/2}(x_2)\right|\exp[-i|w(x_2)|], \tag{3.12.1}$$

where for the sake of simplicity we have written A and B instead of $A(x_2)$ and $B(x_2)$, respectively. The wave function at the point x_1 in the classically forbidden region is then given by (2.4.23), which we write in the slightly different form

$$\psi(x_1) = \left|q^{-1/2}(x_1)\right|\exp[+|w(x_1)|]\{C(x_1) + D(x_1)\exp[-2|w(x_1)|]\}, \tag{3.12.2}$$

using the expressions (2.4.29a,b) for $C(x_1)$ and $D(x_1)$. From these expressions we obtain with the aid of the estimates (2.4.8a–c) for α, β and γ and (2.4.6)

$$C(x_1) = A\exp(-i\pi/4) + B\exp(+i\pi/4) + (|A| + |B|)O(\mu), \qquad (3.12.3a)$$

$$|D(x_1)|\exp[-2|w(x_1)|] = (|A| + |B|)O(\mu). \qquad (3.12.3b)$$

With the requirement that

$$|A\exp(-i\pi/4) + B\exp(+i\pi/4)| \gg (|A| + |B|)\mu, \qquad (3.12.4)$$

(3.12.2) and (3.12.3a,b) yield approximately

$$\psi(x_1) = [A\exp(-i\pi/4) + B\exp(+i\pi/4)]|q^{-1/2}(x_1)|\exp[+|w(x_1)|]. \qquad (3.12.5)$$

From (3.12.1) and (3.12.5) we then obtain the connection formula

$$A|q^{-1/2}(x)|\exp(+i|w(x)|) + B|q^{-1/2}(x)|\exp(-i|w(x)|)$$
$$\rightarrow [A\exp(-i\pi/4) + B\exp(+i\pi/4)]|q^{-1/2}(x)|\exp[+|w(x)|], \qquad (3.12.6)$$

which according to (3.12.4) is valid when $|A\exp(-i\pi/4) + B\exp(+i\pi/4)|$ is not too small compared to $(|A| + |B|)$. This connection formula can be obtained as a special case of the connection formula (3.9.6).

When $A = \exp(+i\pi/4)$, $B = 0$ and $A = 0$, $B = \exp(-i\pi/4)$, respectively, condition (3.12.4) is fulfilled when $\mu \ll 1$, and from (3.12.6) we obtain the connection formula

$$|q^{-1/2}(x)|\exp\{\pm i[|w(x)| + \pi/4]\} \rightarrow |q^{-1/2}(x)|\exp[+|w(x)|], \qquad (3.12.7)$$

where the left-hand side is a complex function, the imaginary part of which gives rise to a contribution to an omitted insignificant term on the right-hand side of (3.12.7).

Putting

$$A = \tfrac{1}{2}\exp[+i(\delta - \pi/4)], \qquad (3.12.8a)$$

$$B = \tfrac{1}{2}\exp[-i(\delta - \pi/4)] \qquad (3.12.8b)$$

instead in (3.12.6), we obtain the connection formula

$$|q^{-1/2}(x)|\cos[|w(x)| + \delta - \pi/4] \rightarrow \sin\delta|q^{-1/2}(x)|\exp[+|w(x)|], \qquad (3.12.9)$$

which according to (3.12.4) and (3.12.8a,b) is valid when $|\sin\delta| \gg \mu$, i.e., when δ is not too close to an integer multiple of π; cf. the connection formula (3.10.10′). Equation (3.20.7b) illustrates what happens when δ is too close to an integer multiple of π. For later discussions we also give explicitly the particular connection formula that one obtains from (3.12.9) by putting $\delta = \pi/2$:

$$\left|q^{-1/2}(x)\right| \cos[|w(x)| + \pi/4] \to \left|q^{-1/2}(x)\right| \exp[|w(x)|]. \qquad (3.12.9')$$

The connection formula (3.12.9), which is a particular case of (3.9.6), was first given by Jeffreys (1925) for the first-order WKB approximation and a particular value of δ (although not in quite the correct form) and by N. Fröman (1970) for an arbitrary order of the phase-integral approximation under the assumption that $Q^2(z) = R(z)$, but according to the derivation in the present section this assumption is not necessary. It is seen that the connection formula has the same form for every order of the phase-integral approximation, but it was not clear that this should be the case until the connection formula had actually been derived for an arbitrary order of approximation. The one-directional nature of the connection formula (3.12.9') has been discussed in detail and even confirmed numerically by N. Fröman (1966a); see eq. (48b) and figs. 8a and 8b in that paper.

3.13 One-directional nature of the connection formulas for tracing a phase-integral wave function from a classically allowed to a classically forbidden region

The connection formulas in Section 3.12 are one-directional. Demonstrate this fact for the connection formula (3.12.9) by simple argumentation, based on the results in Section 3.6, as well as by direct calculation using the parameterized formulas for the F-matrix in Section 2.4.

Solution. The wave function traced by means of the connection formula (3.12.9) is represented in the classically forbidden region by a phase-integral function that decreases exponentially in the direction towards the turning point. Thus, the properties of the phase-integral approximation along a Stokes line, analysed in Section 3.6, already imply that the connection formula (3.12.9) is one-directional. This is obvious, since in (3.12.9) δ appears in the amplitude on the right-hand side, but in the phase on the left-hand side.

To demonstrate explicitly the one-directional nature of the connection formula (3.12.9) by means of the parameterized wave function formulas presented in Section 2.4 we put $C(x_1) = \sin\delta$ and $D(x_1) = 0$, and obtain from (2.4.23)

$$\psi(x_1) = \sin\delta \left|q^{-1/2}(x_1)\right| \exp[+|w(x_1)|] \qquad (3.13.1)$$

and from (2.4.26) and (2.4.28a,b)

$$\psi(x_2) = \sin\delta\left(\frac{1}{2\alpha} + i\alpha\gamma\right)\exp[+i(\pi/4 - \beta)]|q^{-1/2}(x_2)|\exp[+i|w(x_2)|]$$

$$+ \sin\delta\left(\frac{1}{2\alpha} - i\alpha\gamma\right)\exp[-i(\pi/4 - \beta)]|q^{-1/2}(x_2)|\exp[-i|w(x_2)|]$$

$$= \sin\delta\left(\frac{1}{\alpha^2} + 4\alpha^2\gamma^2\right)^{1/2}|q^{-1/2}(x_2)|\cos[|w(x_2)| + \pi/4 - \beta + \arctan(2\alpha^2\gamma)],$$

$$-\pi/2 < \arctan(2\alpha^2\gamma) < +\pi/2. \quad (3.13.2)$$

By means of the estimates (2.4.8a,b) we obtain from (3.13.2) the approximate formula

$$\psi(x_2) \approx \sin\delta(1 + 4\gamma^2)^{1/2}|q^{-1/2}(x_2)|\cos[|w(x_2)| + \pi/4 + \arctan(2\gamma)],$$

$$-\pi/2 < \arctan(2\gamma) < +\pi/2. \quad (3.13.3)$$

Since the absolute value of γ is in general very large compared with unity, we realize that in general neither the amplitude nor the phase of the expression (3.13.3) agrees even roughly with the corresponding quantity on the left-hand side of the connection formula (3.12.9). Agreement occurs only if accidentally $\delta - \pi/4 = \pi/4 + \arctan(2\gamma)$, $-\pi/2 < \arctan(2\gamma) < +\pi/2$, which implies that $(1 + 4\gamma^2)^{1/2}\sin\delta = 1$. However, this result of our analysis cannot be used in practice, since γ is unknown. Thus, the connection formula (3.12.9) is one-directional for all practical purposes.

Although the absolute value of the parameter γ is in general large compared with unity, there sometimes exists an exceptional point in the classically forbidden region where γ passes steeply through zero and thus changes sign. In this situation we obtain from (3.13.3) the approximate formulas

$$\psi(x_2) = 2\sin\delta|\gamma q^{-1/2}(x_2)|\cos[|w(x_2)| + 3\pi/4] \quad \text{when } \gamma \gg 1/2, \quad (3.13.4a)$$

$$\psi(x_2) = \sin\delta|q^{-1/2}(x_2)|\cos[|w(x_2)| + \pi/4] \quad \text{when } |\gamma| \ll 1/2, \quad (3.13.4b)$$

$$\psi(x_2) = 2\sin\delta|\gamma q^{-1/2}(x_2)|\cos[|w(x_2)| - \pi/4] \quad \text{when } \gamma \ll -1/2. \quad (3.13.4c)$$

We remark that by replacing $\sin\delta$ by $C(x_1)$ in (3.13.4a,c), one obtains (3.10.13). When $|\gamma| \gg 1/2$ the phase of the cosine is according to (3.13.4a,c) known modulo π, but the amplitude is unknown and large compared with unity. As γ passes through zero, the phase changes rapidly by π; for a concrete illustration of how rapidly this change occurs we refer the interested reader to eqs. (48a,b) and figs. 8a and 8b in N. Fröman (1966a) along with (2.4.2b) and (2.4.7c''). The x_1-interval, in which (3.13.4b) is valid, is very small, and its position is unknown. Thus, (3.13.4b) is of no practical importance and cannot justify the use of the one-directional connection formula (3.12.9′) in the forbidden direction.

3.14 Value at the turning point of the wave function associated with the connection formula for tracing a phase-integral wave function from the classically forbidden to the classically allowed region

Use the Airy differential equation as the comparison equation and derive, in the first-order approximation, the connection formula for tracing the wave function from the classically forbidden to the classically allowed region. Then calculate the value of this wave function at the turning point.

Solution. The differential equation is

$$d^2\psi/dz^2 + R(z)\psi = 0. \tag{3.14.1}$$

Without specifying the base function $Q(z)$, we assume that $Q^2(z) > 0$ for $z < t$ and $Q^2(z) < 0$ for $z > t$, i.e., that the classically allowed region lies to the left and the classically forbidden region lies to the right of t. The variable z is assumed to lie on the real axis or in the upper half of the complex z-plane, and the phase of $Q(z)$ is chosen such that $Q(z) = |Q(z)|$ for $z < t$ and $Q(z) = \exp(-i\pi/2)|Q(z)|$ for $z > t$.

The comparison function appropriate for the solution of the present problem is expressed in terms of the Airy function $\text{Ai}(\varphi)$, which satisfies the differential equation

$$d^2\text{Ai}(\varphi)/d\varphi^2 - \varphi\text{Ai}(\varphi) = 0, \tag{3.14.2}$$

with $\varphi < 0$ corresponding to the classically allowed region $z < t$, where $Q^2(z) > 0$, and $\varphi > 0$ corresponding to the classically forbidden region, where $Q^2(z) < 0$. To express φ in terms of z we use the relation (cf. eq. (2.2.20) in chapter 2 of Fröman and Fröman (1996))

$$\int_t^z Q(z)dz = \int_0^\varphi [\exp(-i\pi)\varphi]^{1/2}d\varphi = \tfrac{2}{3}\exp(-i\pi/2)\varphi^{3/2}, \tag{3.14.3}$$

which means that the turning point t in the z-plane is mapped into the point $\varphi = 0$ in the φ-plane. From (3.14.3) we obtain the formula

$$\varphi = \left[\tfrac{3}{2}\exp(i\pi/2)\int_t^z Q(z)dz\right]^{2/3}, \tag{3.14.4}$$

which yields φ explicitly as a function of z, when the base function $Q(z)$ has been chosen. We write the conveniently normalized comparison equation solution of (3.14.1) as (cf. eq. (2.2.3) in chapter 2 of Fröman and Fröman (1996))

$$\psi = 2\pi^{1/2}(d\varphi/dz)^{-1/2}\text{Ai}(\varphi).\tag{3.14.5}$$

By differentiating (3.14.3) we find that

$$d\varphi/dz = \exp(i\pi/2)Q(z)\varphi^{-1/2} = [\exp(i\pi)Q^2(z)/\varphi]^{1/2},\tag{3.14.6}$$

and with the aid of (3.14.6) we can write (3.14.5) as

$$\psi = 2\pi^{1/2}[\exp(-i\pi)\varphi/Q^2(z)]^{1/4}\text{Ai}(\varphi).\tag{3.14.7}$$

When $z > t$ we have $Q^2(z) = |Q^2(z)|\exp(-i\pi)$ and $\varphi > 0$, and we write (3.14.7) as

$$\psi = 2\pi^{1/2}|Q^{-1/2}(z)|\,\varphi^{1/4}\text{Ai}(\varphi)\tag{3.14.8}$$

and use the asymptotic formula

$$\text{Ai}(\varphi) \sim \tfrac{1}{2}\pi^{-1/2}\varphi^{-1/4}\exp\!\left(-\tfrac{2}{3}\varphi^{3/2}\right), \quad \varphi \to +\infty,\tag{3.14.9}$$

and (3.14.3). Thus we obtain the approximate formula

$$\psi = |Q^{-1/2}(z)|\exp\left[-\left|\int_t^z Q(z)dz\right|\right], \quad z > t.\tag{3.14.10}$$

When $z < t$ we have $Q^2(z) = |Q^2(z)|$ and $\varphi = |\varphi|\exp(i\pi)$. Writing (3.14.7) as

$$\psi = 2\pi^{1/2}|Q^{-1/2}(z)|[\exp(-i\pi)\varphi]^{1/4}\text{Ai}(\varphi)\tag{3.14.11}$$

and using the asymptotic formula

$$\text{Ai}(\varphi) \sim \pi^{-1/2}[\exp(-i\pi)\varphi]^{-1/4}\cos\{\tfrac{2}{3}[\exp(-i\pi)\varphi]^{3/2} - \pi/4\},$$
$$\varphi = |\varphi|\exp(i\pi) \to -\infty,\tag{3.14.12}$$

along with (3.14.3), we obtain the approximate formula

$$\psi = 2|Q^{-1/2}(z)|\cos\left[\left|\int_t^z Q(z)dz\right| - \pi/4\right], \quad z < t.\tag{3.14.13}$$

Now we write (3.14.10) and (3.14.13) in the form of the connection formula

$$|Q^{-1/2}(z)|\exp\left[-\left|\int_t^z Q(z)dz\right|\right] \rightarrow 2|Q^{-1/2}(z)|\cos\left[\left|\int_t^z Q(z)dz\right| - \pi/4\right],$$

(3.14.14)

which is one-directional, although this is not seen in the present derivation.

When z lies close to the turning point t we have approximately

$$Q^2(z) = [dQ^2(z)/dz]_{z=t}(z-t) = \exp(-i\pi)|dQ^2(z)/dz|_{z=t}(z-t),$$

(3.14.15)

and hence

$$Q(z) = \exp(-i\pi/2)|dQ^2(z)/dz|_{z=t}^{1/2}(z-t)^{1/2},$$

(3.14.16)

$$\int_t^z Q(z)dz = \frac{2}{3}\exp(-i\pi/2)|dQ^2(z)/dz|_{z=t}^{1/2}(z-t)^{3/2}.$$

(3.14.17)

Insertion of (3.14.17) into (3.14.4) yields

$$\varphi = |dQ^2(z)/dz|_{z=t}^{1/3}(z-t),$$

(3.14.18)

and (3.14.15) and (3.14.18) yield the approximate formula

$$[\exp(-i\pi)\varphi/Q^2(z)]^{1/4} = |dQ^2(z)/dz|_{z=t}^{-1/6},$$

(3.14.19)

which is valid in the limit when $z \rightarrow t$. In this limit, i.e., for $\varphi = 0$, the value of the Airy function is

$$\mathrm{Ai}(0) = \frac{1}{3^{2/3}\Gamma(2/3)}.$$

(3.14.20)

Inserting (3.14.19) and (3.14.20) into (3.14.7) with $z = t$, we obtain the approximate formula

$$\psi(t) = \frac{2\pi^{1/2}|dQ^2(z)/dz|_{z=t}^{-1/6}}{3^{2/3}\Gamma(2/3)} = 1.259|dQ^2(z)/dz|_{z=t}^{-1/6}$$

(3.14.21)

for the value of the wave function at the turning point t.

In the treatment above we have used the linear approximation (3.14.15) of $Q^2(z)$ only in an arbitrarily small interval containing t. One can derive (3.14.21) in a simpler way by introducing the more restrictive assumptions that $Q^2(z) = R(z)$ and that $R(z)$ is approximately linear in an interval around t that is so large that the asymptotic formulas for the Airy function can be used in the outer parts of this interval. In this interval we thus have approximately

$$R(z) = Q^2(z) = [dQ^2(z)/dz]_{z=t}(z-t),$$

(3.14.22)

and the differential equation (3.14.1) is there approximately replaced by the differential equation

$$d^2\psi/dz^2 + [dQ^2(z)/dz]_{z=t}(z-t)\psi = 0, \tag{3.14.23}$$

which we write as

$$d^2\psi/d\varphi^2 - \varphi\psi = 0, \tag{3.14.24}$$

where

$$\varphi = [-dQ^2(z)/dz]_{z=t}^{1/3}(z-t) = |dQ^2(z)/dz|_{z=t}^{1/3}(z-t) \tag{3.14.25}$$

in agreement with (3.14.18). The function

$$\psi = 2\pi^{1/2}|dQ^2(z)/dz|_{z=t}^{-1/6}\text{Ai}(\varphi) \tag{3.14.26}$$

is a particular solution of (3.14.24) with an asymptoic behaviour as $\varphi \to +\infty$ that is convenient for our present purpose. According to the assumption stated above, the region in which the approximate differential equation (3.14.23) and hence its approximate solution (3.14.26) can be used reaches from the turning point t so far into the classically forbidden and classically allowed regions that the asymptotic formulas (3.14.9) and (3.14.12) are valid. When $\varphi \gg 1$ we obtain from (3.14.26) with the aid of (3.14.9), (3.14.25) and (3.14.22)

$$\psi \sim \frac{|dQ^2(z)/dz|_{z=t}^{-1/6}}{\varphi^{1/4}} \exp\left(-\frac{2}{3}\varphi^{3/2}\right)$$

$$= [|dQ^2(z)/dz|_{z=t}(z-t)]^{-1/4}\exp\left[-\frac{2}{3}|dQ^2(z)/dz|_{z=t}^{1/2}[(z-t)]^{3/2}\right]$$

$$= |Q^{-1/2}(z)|\exp\left(-\left|\int_t^z Q(z)dz\right|\right), \quad z > t, \tag{3.14.27}$$

in agreement with (3.14.10). When $\varphi \ll -1$ we obtain from (3.14.26) with the aid of (3.14.12), (3.14.25) and (3.14.22)

$$\psi \sim \frac{2|dQ^2(z)/dz|_{z=t}^{-1/6}}{(\exp(-i\pi)\varphi)^{1/4}} \cos\left\{\frac{2}{3}[\exp(-i\pi)\varphi]^{3/2} - \pi/4\right\}$$

$$= \frac{2}{[|dQ^2(z)/dz|_{z=t}(t-z)]^{1/4}} \cos\left[\frac{2}{3}|dQ^2(z)/dz|_{z=t}^{1/2}(t-z)^{3/2} - \pi/4\right]$$

$$= 2|Q^{-1/2}(z)|\cos\left[\left|\int_t^z Q(z)dz\right| - \pi/4\right] \tag{3.14.28}$$

in agreement with (3.14.13). For $z = t$ we obtain from (3.14.26) the formula

$$\psi(t) = 2\pi^{1/2} |d Q^2(z)/dz|_{z=t}^{-1/6} \text{Ai}(0),\tag{3.14.29}$$

which along with (3.14.20) is in agreement with (3.14.21). The accuracy of (3.14.21) is checked in Section 3.16 for a particular differential equation.

3.15 Value at the turning point of the wave function associated with a connection formula for tracing the phase-integral wave function from the classically allowed to the classically forbidden region

Use the Airy differential equation as the comparison equation and derive, in the first-order approximation, the connection formula (3.12.9′) for tracing the wave function from the classically allowed to the classically forbidden region. Then calculate the value of this wave function at the turning point.

Solution. The solution of this problem is similar to the solution in Section 3.14, and therefore we only sketch the procedure. The comparison function appropriate for the solution of the present problem is expressed in terms of the Airy function $\text{Bi}(\varphi)$, which is a solution of the differential equation

$$d^2\text{Bi}(\varphi)/d\varphi^2 - \varphi\text{Bi}(\varphi) = 0.\tag{3.15.1}$$

The conveniently normalized comparison equation solution is

$$\bar{\psi} = \pi^{1/2}(d\varphi/dz)^{-1/2} \text{Bi}(\varphi).\tag{3.15.2}$$

When $z < t$ we use the asymptotic formula

$$\text{Bi}(\varphi) \sim \pi^{-1/2}[\exp(-i\pi)\varphi]^{-1/4} \cos\{\tfrac{2}{3}[\exp(-i\pi)\varphi]^{3/2} + \pi/4\}\tag{3.15.3}$$

and obtain from (3.15.2), on recalling (3.14.3) and (3.14.6),

$$\bar{\psi} \sim |Q^{-1/2}(z)| \cos\left[\left|\int_t^z Q(z)dz\right| + \pi/4\right].\tag{3.15.4}$$

When $z > t$ we use the asymptotic formula

$$\text{Bi}(\varphi) \sim \pi^{-1/2}\varphi^{-1/4}\exp(\tfrac{2}{3}\varphi^{3/2})\tag{3.15.5}$$

and obtain in a manner analogous to that above

$$\bar{\psi} \sim |Q^{-1/2}(z)|\exp\left[\left|\int_t^z Q(z)dz\right|\right].\tag{3.15.6}$$

Now we write (3.15.4) and (3.15.6) in the form of the connection formula

$$
|Q^{-1/2}(z)|\cos\left[\left|\int_t^z Q(z)dz\right| + \pi/4\right] \to |Q^{-1/2}(z)|\exp\left[\left|\int_t^z Q(z)dz\right|\right],
$$

(3.15.7)

which is one-directional, although this is not seen in the present derivation.

For $z = t$ we use the formula

$$
\mathrm{Bi}(0) = \frac{1}{3^{1/6}\Gamma(2/3)}
$$

(3.15.8)

and obtain from (3.15.2) using (3.14.6) and (3.14.19)

$$
\bar{\psi}(t) = \frac{\pi^{1/2}|d\,Q^2(z)/dz|_{z=t}^{-1/6}}{3^{1/6}\Gamma(2/3)} = 1.090|d\,Q^2(z)/dz|_{z=t}^{-1/6}.
$$

(3.15.9)

The accuracy of (3.15.9) is checked in Section 3.16 for a particular differential equation.

For the particular case in which $Q^2(z) = R(z)$ and $R(z)$ can be approximated by a linear function of z in a sufficiently large interval on both sides of t, one can derive (3.15.9) in a simpler way, which is analogous to the corresponding derivation described at the end of Section 3.14.

3.16 Illustration of the accuracy of the approximate formulas for the value of the wave function at a turning point

Apply the formulas for the value of the wave function at the turning point, derived in Sections 3.14 and 3.15, to solutions of the radial Schrödinger equation for a free particle, and compare those results with the results obtained from exact solutions of the differential equation in question.

Solution. The radial Schrödinger equation for a free particle with the angular momentum quantum number l is, with obvious notation,

$$
d^2\psi/dz^2 + R(z)\psi = 0,
$$

(3.16.1a)

$$
R(z) = 1 - l(l+1)/z^2,
$$

(3.16.1b)

$$
z = kr = (2mE/\hbar^2)^{1/2}r.
$$

(3.16.1c)

With the usual choice (2.2.12a) of the square of the base function for the radial Schrödinger equation we have

$$
Q^2(z) = R(z) - 1/(4z^2) = 1 - t^2/z^2,
$$

(3.16.2a)

$$
t = l + 1/2.
$$

(3.16.2b)

Connection from the classically forbidden to the classically allowed region. First we consider the phase-integral solution on the left-hand side of the connection formula (3.14.14), i.e., the phase-integral function that tends to zero at the origin, and the corresponding exact solution. When $z < t$ we obtain on using (3.16.2a)

$$\left| \int_t^z Q(z) dz \right| = \int_z^t (t^2/z^2 - 1)^{1/2} dz = t \ln \frac{t + (t^2 - z^2)^{1/2}}{z} - (t^2 - z^2)^{1/2}$$

$$\approx t \ln(2t/z) - t, \quad z (>0) \text{ sufficiently small.} \tag{3.16.3}$$

Close to the origin we thus have

$$|Q^{-1/2}(z)| \exp\left[-\left| \int_t^z Q(z) dz \right| \right] \approx 2^{1/2} \exp(t)[z/(2t)]^{t+1/2}$$

$$= 2^{1/2} \exp(l + 1/2)[z/(2l + 1)]^{l+1}. \tag{3.16.4}$$

We shall now verify that the exact solution of (3.16.1a,b) which for $z < t$ and $t = l + 1/2$ sufficiently large is given by the left-hand side of (3.16.4) and hence for sufficiently small values of z is given by the right-hand side of (3.16.4), is

$$\psi = 2z\, j_l(z), \tag{3.16.5}$$

where $j_l(z)$ is a spherical Bessel function. When the absolute value of z is sufficiently small, we have for the function (3.16.5) the approximate formula (cf. (3.16.2b))

$$\psi \approx \frac{2^{l+1} \Gamma(l + 1)}{\Gamma(2l + 2)} z^{l+1}. \tag{3.16.6}$$

Using the asymptotic formula (5) on page 32 in Luke (1969), i.e.,

$$\ln \Gamma(z + 1/2) \sim z \ln z - z + \ln(2\pi)^{1/2}, \quad |\arg(z)| \le \pi - \varepsilon, \quad \varepsilon > 0,$$

we obtain

$$\frac{2^{l+1} \Gamma(l + 1)}{\Gamma(2l + 2)} \approx \frac{2^{1/2} \exp(l + 1/2)}{(2l + 1)^{l+1}}, \tag{3.16.7}$$

and inserting (3.16.7) into (3.16.6), we obtain

$$\psi \approx 2^{1/2} \exp(l + 1/2)[z/(2l + 1)]^{l+1}, \quad z \to 0. \tag{3.16.8}$$

By comparing the right-hand sides of (3.16.4) and (3.16.8), we can verify our assertion. The value at t of the wave function that close to the origin is represented by the left-hand side of (3.16.4) is thus (cf. (3.16.5) and (3.16.2b))

$$\psi(t) = 2t\, j_l(t) = (2l + 1) j_l(l + 1/2). \tag{3.16.9}$$

Table 3.16.1. *Illustration of the accuracy of (3.16.10)*.

l	$\psi(t)$ obtained from (3.16.9)	$\psi(t)$ obtained from (3.16.10)
0	0.959	0.999
1	1.189	1.200
2	1.3003	1.3062
3	1.3777	1.3816
4	1.4378	1.4407
5	1.4874	1.4897
6	1.5299	1.5317
7	1.5672	1.5687
8	1.6005	1.6018
9	1.6306	1.6317
10	1.6582	1.6592

The corresponding approximate value obtained from (3.14.21) and (3.16.2a,b) is

$$\psi(t) = \frac{2\pi^{1/2}[(2l+1)/4]^{1/6}}{3^{2/3}\Gamma(2/3)} = 1.259[(2l+1)/4]^{1/6}. \qquad (3.16.10)$$

Table 3.16.1 compares the values of $\psi(t)$ obtained from (3.16.9), which is based on the exact solution of the differential equation, with approximate values obtained from (3.16.10).

Connection from the classically allowed to the classically forbidden region. Now we consider the phase-integral solution on the left-hand side of the connection formula (3.15.7) and the corresponding exact solution. When $z > t$ we obtain on using (3.16.2a,b)

$$\left| \int_t^z Q(z)dz \right| = \int_t^z (1 - t^2/z^2)^{1/2}dz = (z^2 - t^2)^{1/2} + t \arctan\frac{t}{(z^2 - t^2)^{1/2}} - t\pi/2$$

$$\approx z - t\pi/2 = z - (l+1/2)\pi/2, \quad z(>t = l+1/2) \text{ sufficiently large}, \qquad (3.16.11)$$

and hence

$$\left| Q^{-1/2}(z) \right| \cos \left[\left| \int_t^z Q(z)dz \right| + \pi/4 \right] \approx \cos(z - l\pi/2). \qquad (3.16.12)$$

Table 3.16.2. *Illustration of the accuracy of (3.16.15).*

l	$\bar{\psi}(t)$ obtained from (3.16.14)	$\bar{\psi}(t)$ obtained from (3.16.15)
0	0.878	0.865
1	1.045	1.039
2	1.135	1.131
3	1.199	1.196
4	1.2496	1.2477
5	1.2916	1.2909
6	1.3278	1.3265
7	1.3596	1.3585
8	1.3881	1.3872
9	1.4140	1.4131
10	1.4377	1.4369

The exact solution, which as $z \to +\infty$ is given by the expression on the right-hand side of (3.16.12), is

$$\bar{\psi}(z) = -zy_l(z), \qquad (3.16.13)$$

where $y_l(z)$ is a spherical Bessel function. The exact value at t of this solution is thus (cf. (3.16.2b))

$$\bar{\psi}(t) = -ty_l(t) = -(l + 1/2)y_l(l + 1/2). \qquad (3.16.14)$$

The corresponding approximate value obtained from (3.15.9) and (3.16.2a,b) is

$$\bar{\psi}(t) = \frac{\pi^{1/2}[(2l + 1)/4]^{1/6}}{3^{1/6}\Gamma(2/3)} = 1.090[(2l + 1)/4]^{1/6}. \qquad (3.16.15)$$

Table 3.16.2 compares the exact values of $\bar{\psi}(t)$ obtained from (3.16.14) with approximate values obtained from (3.16.15).

3.17 Expressions for the a-coefficients associated with the Airy functions

Use (2.3.36a,b), (2.3.37a,b) and well-known asymptotic formulas for the Airy functions to determine, in the first-order approximation with $Q^2(z) = R(z)$, expressions for the a-coefficients associated with the Airy functions. The final results will be used in Section 3.18 to illuminate the estimates (2.4.8a–c) for the parameters α, β and γ in the parameterized F-matrix. They will also be used in Section 3.19, which prepares the ground for the discussion of the connection formulas in Section 3.20.

Solution. The Airy functions $\mathrm{Ai}(z)$ and $\mathrm{Bi}(z)$ are linearly independent solutions of the differential equation (2.1.1) with $R(z) = -z$, i.e.,

$$d^2\psi/dz^2 - z\psi = 0. \tag{3.17.1}$$

Recalling (2.3.4a,b), we write

$$\mathrm{Ai}(z) = a_1(z)f_1(z) + a_2(z)f_2(z), \tag{3.17.2a}$$

$$\mathrm{Ai}'(z) = a_1(z)f_1'(z) + a_2(z)f_2'(z), \tag{3.17.2b}$$

$$\mathrm{Bi}(z) = \bar{a}_1(z)f_1(z) + \bar{a}_2(z)f_2(z), \tag{3.17.3a}$$

$$\mathrm{Bi}'(z) = \bar{a}_1(z)f_1'(z) + \bar{a}_2(z)f_2'(z), \tag{3.17.3b}$$

where in the first-order approximation we have according to (2.3.1a–c) with $Z = 0$

$$f_1(z) = Q^{-1/2}(z)\exp[+i\zeta(z)], \tag{3.17.4a}$$

$$f_2(z) = Q^{-1/2}(z)\exp[-i\zeta(z)], \tag{3.17.4b}$$

$$\zeta(z) = \int_0^z Q(z)dz. \tag{3.17.4c}$$

Putting $\psi(z) = \mathrm{Ai}(z)$ in (2.3.36a,b) and $\bar{\psi}(z) = \mathrm{Bi}(z)$ in (2.3.37a,b), we obtain

$$a_1(z) = -\frac{1}{2}i[f_2(z)\mathrm{Ai}'(z) - f_2'(z)\mathrm{Ai}(z)], \tag{3.17.5a}$$

$$a_2(z) = +\frac{1}{2}i[f_1(z)\mathrm{Ai}'(z) - f_1'(z)\mathrm{Ai}(z)], \tag{3.17.5b}$$

$$\bar{a}_1(z) = -\frac{1}{2}i[f_2(z)\mathrm{Bi}'(z) - f_2'(z)\mathrm{Bi}(z)], \tag{3.17.6a}$$

$$\bar{a}_2(z) = +\frac{1}{2}i[f_1(z)\mathrm{Bi}'(z) - f_1'(z)\mathrm{Bi}(z)]. \tag{3.17.6b}$$

Let x_1 be a point with the argument zero in the classically forbidden region ($z > 0$), and let x_2 be a point with the argument $+\pi$ in the classically allowed region ($z < 0$). With

$$Q^2(z) = R(z) = -z = \exp(-i\pi)z = |z|\exp[i(\arg z - \pi)], \quad 0 \le \arg z \le \pi,$$

we have (cf. (3.17.4c))

$$Q^{-1/2}(x_1) = \exp(i\pi/4)x_1^{-1/4} = \exp(i\pi/4)|Q^{-1/2}(x_1)|, \tag{3.17.7a}$$

$$\zeta(x_1) = -\frac{2}{3}ix_1^{3/2} = -i|\zeta(x_1)|, \tag{3.17.7b}$$

$$Q^{-1/2}(x_2) = [\exp(-i\pi)x_2]^{-1/4} = |x_2|^{-1/4} = \left|Q^{-1/2}(x_2)\right|, \quad (3.17.8a)$$

$$\zeta(x_2) = -\frac{2}{3}[\exp(-i\pi)x_2]^{3/2} = -\frac{2}{3}|x_2|^{3/2} = -|\zeta(x_2)|. \quad (3.17.8b)$$

Using (3.17.4a,b), (3.17.7a,b) and (3.17.8a,b), we obtain

$$f_1(x_1) = \exp(i\pi/4)x_1^{-1/4}\exp\left(\tfrac{2}{3}x_1^{3/2}\right), \quad (3.17.9a)$$

$$f_2(x_1) = \exp(i\pi/4)x_1^{-1/4}\exp\left(-\tfrac{2}{3}x_1^{3/2}\right), \quad (3.17.9b)$$

$$f_1(x_2) = [\exp(-i\pi)x_2]^{-1/4}\exp\left\{-\tfrac{2}{3}i[\exp(-i\pi)x_2]^{3/2}\right\}, \quad (3.17.10a)$$

$$f_2(x_2) = [\exp(-i\pi)x_2]^{-1/4}\exp\left\{+\tfrac{2}{3}i[\exp(-i\pi)x_2]^{3/2}\right\}. \quad (3.17.10b)$$

According to formulas on pp. 103, 105 and 106 in Copson (1965) and our formulas (3.17.9a,b), (3.17.10a,b), (3.17.7a,b) and (3.17.8a,b) the asymptotic expansions for the Airy functions at the points $x_1(\to +\infty)$ and $x_2(\to -\infty)$ are

$$\mathrm{Ai}(x_1) \sim \frac{\exp(-i\pi/4)}{2\pi}f_2(x_1)\sum_{n=0}^{\infty}\frac{\Gamma(3n+1/2)}{(2n)!\left(-9x_1^{3/2}\right)^n} \quad (3.17.11a)$$

$$\approx \frac{\exp(-i\pi/4)}{2\pi^{1/2}}f_2(x_1) = \frac{1}{2\pi^{1/2}}\left|Q^{-1/2}(x_1)\right|\exp[-|\zeta(x_1)|], \quad (3.17.11a')$$

$$\mathrm{Bi}(x_1) \sim \frac{\exp(-i\pi/4)}{\pi}f_1(x_1)\sum_{n=0}^{\infty}\frac{\Gamma(3n+1/2)}{(2n)!\left(9x_1^{3/2}\right)^n} \quad (3.17.11b)$$

$$\approx \pi^{-1/2}\exp(-i\pi/4)f_1(x_1) = \pi^{-1/2}\left|Q^{-1/2}(x_1)\right|\exp[+|\zeta(x_1)|],$$
$$(3.17.11b')$$

$$\mathrm{Ai}(x_2) \sim \frac{\exp(-i\pi/4)}{2\pi}f_2(x_2)\sum_{n=0}^{\infty}\frac{\Gamma(3n+1/2)}{(9i)^n(2n)!\,[x_2\exp(-i\pi)]^{3n/2}}$$

$$+ \frac{\exp(+i\pi/4)}{2\pi}f_1(x_2)\sum_{n=0}^{\infty}\frac{\Gamma(3n+1/2)}{(-9i)^n\,(2n)!\,[x_2\exp(-i\pi)]^{3n/2}} \quad (3.17.12a)$$

$$\approx \frac{\exp(-i\pi/4)}{2\pi^{1/2}}f_2(x_2) + \frac{\exp(+i\pi/4)}{2\pi^{1/2}}f_1(x_2)$$

$$= \frac{1}{\pi^{1/2}}\left|Q^{-1/2}(x_2)\right|\cos[|\zeta(x_2)| - \pi/4], \quad (3.17.12a')$$

$$\mathrm{Bi}(x_2) \sim \frac{\exp(+i\pi/4)}{2\pi}f_2(x_2)\sum_{n=0}^{\infty}\frac{\Gamma(3n+1/2)}{(9i)^n(2n)!\,[x_2\exp(-i\pi)]^{3n/2}}$$

$$+ \frac{\exp(-i\pi/4)}{2\pi}f_1(x_2)\sum_{n=0}^{\infty}\frac{\Gamma(3n+1/2)}{(-9i)^n(2n)!\,[x_2\exp(-i\pi)]^{3n/2}} \quad (3.17.12b)$$

$$\approx \frac{\exp(+i\pi/4)}{2\pi^{1/2}} f_2(x_2) + \frac{\exp(-i\pi/4)}{2\pi^{1/2}} f_1(x_2)$$

$$= \frac{1}{\pi^{1/2}} |Q^{-1/2}(x_2)| \cos[|\zeta(x_2)| + \pi/4]. \qquad (3.17.12b')$$

In (3.17.12b) we have corrected a misprint on p. 106 in Copson (1965); a corresponding misprint appears also in the asymptotic series for Ai($-z$) on p. 105 in the same book. Using (2.3.2) and (3.17.4a,b)–(3.17.12a,b,b'), we obtain after straightforward calculations

$$a_1(x_1) \sim -\frac{\exp(i\pi/4)}{4\pi} [f_2(x_1)]^2 \frac{d}{dx_1} \sum_{n=0}^{\infty} \frac{\Gamma(3n+1/2)}{(2n)!(-9x_1^{3/2})^n}$$

$$= \frac{27\exp(-i\pi/4)}{16\pi} \exp[-2i\zeta(x_1)] \sum_{n=1}^{\infty} \frac{\Gamma(3n+1/2)}{(2n-1)!\left[-\dfrac{27i}{2}\zeta(x_1)\right]^{n+1}}$$

$$= \frac{5\exp(-i\pi/4)}{288\pi^{1/2}|\zeta(x_1)|^2} \exp[-2|\zeta(x_1)|](1+\cdots), \qquad (3.17.13a)$$

$$a_2(x_1) \sim \frac{\exp(-i\pi/4)}{2\pi} \sum_{n=0}^{\infty} \frac{\Gamma(3n+1/2)}{(2n)!\left(-9x_1^{3/2}\right)^n}$$

$$+ \frac{\exp(+i\pi/4)}{4\pi} f_1(x_1)f_2(x_1)\frac{d}{dx_1} \sum_{n=0}^{\infty} \frac{\Gamma(3n+1/2)}{(2n)!\left(-9x_1^{3/2}\right)^n}$$

$$= \frac{\exp(-i\pi/4)}{2\pi} \sum_{n=0}^{\infty} \frac{\Gamma(3n+1/2)}{(2n)!\left[-\dfrac{27i}{2}\zeta(x_1)\right]^n}$$

$$+ \frac{27\exp(+3i\pi/4)}{16\pi} \sum_{n=1}^{\infty} \frac{\Gamma(3n+1/2)}{(2n-1)!\left[-\dfrac{27i}{2}\zeta(x_1)\right]^{n+1}}$$

$$= \frac{\exp(-i\pi/4)}{2\pi^{1/2}} \left[1 - \frac{5}{72|\zeta(x_1)|} + \frac{25}{10368|\zeta(x_1)|^2} + \cdots\right], \qquad (3.17.13b)$$

$$\bar{a}_1(x_1) \sim \frac{\exp(-i\pi/4)}{\pi} \sum_{n=0}^{\infty} \frac{\Gamma(3n+1/2)}{(2n)!(9x_1^{3/2})^n}$$

$$+ \frac{\exp(-3i\pi/4)}{2\pi} f_1(x_1)f_2(x_1)\frac{d}{dx_1} \sum_{n=0}^{\infty} \frac{\Gamma(3n+1/2)}{(2n)!\left(9x_1^{3/2}\right)^n}$$

$$= \frac{\exp(-i\pi/4)}{\pi} \sum_{n=0}^{\infty} \frac{\Gamma(3n+1/2)}{(2n)! \left[\dfrac{27i}{2}\zeta(x_1)\right]^n}$$

$$+ \frac{27\exp(+3i\pi/4)}{8\pi} \sum_{n=1}^{\infty} \frac{\Gamma(3n+1/2)}{(2n-1)! \left[\dfrac{27i}{2}\zeta(x_1)\right]^{n+1}}$$

$$= \frac{\exp(-i\pi/4)}{\pi^{1/2}} \left[1 + \frac{5}{72\,|\zeta(x_1)|} + \frac{25}{10368\,|\zeta(x_1)|^2} + \cdots\right], \quad (3.17.14a)$$

$$\bar{a}_2(x_1) \sim \frac{\exp(+i\pi/4)}{2\pi}[f_1(x_1)]^2 \frac{d}{dx_1} \sum_{n=0}^{\infty} \frac{\Gamma(3n+1/2)}{(2n)!\,(9x_1^{3/2})^n}$$

$$= \frac{27\exp(-i\pi/4)}{8\pi}\exp[2i\zeta(x_1)] \sum_{n=1}^{\infty} \frac{\Gamma(3n+1/2)}{(2n-1)! \left[\dfrac{27i}{2}\zeta(x_1)\right]^{n+1}}$$

$$= \frac{5\exp(-i\pi/4)}{144\pi^{1/2}\,|\zeta(x_1)|^2}\exp[2\,|\zeta(x_1)|](1 + \cdots), \quad (3.17.14b)$$

$$a_1(x_2) \sim \frac{\exp(+i\pi/4)}{2\pi} \sum_{n=0}^{\infty} \frac{\Gamma(3n+1/2)}{(-9i)^n(2n)!\,[x_2\exp(-i\pi)]^{3n/2}}$$

$$+ \frac{\exp(-i\pi/4)}{4\pi} f_1(x_2)f_2(x_2)\frac{d}{dx_2} \sum_{n=0}^{\infty} \frac{\Gamma(3n+1/2)}{(-9i)^n(2n)!\,[x_2\exp(-i\pi)]^{3n/2}}$$

$$+ \frac{\exp(-3i\pi/4)}{4\pi}[f_2(x_2)]^2 \frac{d}{dx_2} \sum_{n=0}^{\infty} \frac{\Gamma(3n+1/2)}{(9i)^n(2n)!\,[x_2\exp(-i\pi)]^{3n/2}}$$

$$= \frac{\exp(+i\pi/4)}{2\pi} \sum_{n=0}^{\infty} \frac{\Gamma(3n+1/2)}{(2n)! \left[\dfrac{27i}{2}\zeta(x_2)\right]^n}$$

$$+ \frac{27\exp(-3i\pi/4)}{16\pi} \sum_{n=1}^{\infty} \frac{\Gamma(3n+1/2)}{(2n-1)! \left[\dfrac{27i}{2}\zeta(x_2)\right]^{n+1}}$$

$$+ \frac{27\exp(-i\pi/4)}{16\pi}\exp[-2i\zeta(x_2)] \sum_{n=1}^{\infty} \frac{\Gamma(3n+1/2)}{(2n-1)! \left[-\dfrac{27i}{2}\zeta(x_2)\right]^{n+1}}$$

$$= \frac{\exp(+i\pi/4)}{2\pi^{1/2}} \left[1 + \frac{5i}{72\,|\zeta(x_2)|} + \cdots\right], \quad (3.17.15a)$$

$$a_2(x_2) = a_1^*(x_2) \sim \frac{\exp(-i\pi/4)}{2\pi^{1/2}} \left[1 - \frac{5i}{72\,|\zeta(x_2)|} + \cdots \right], \qquad (3.17.15b)$$

$$\bar{a}_1(x_2) \sim \frac{\exp(-i\pi/4)}{2\pi} \sum_{n=0}^{\infty} \frac{\Gamma(3n+1/2)}{(-9i)^n(2n)!\,[x_2\exp(-i\pi)]^{3n/2}}$$

$$+ \frac{\exp(-3i\pi/4)}{4\pi} f_1(x_2) f_2(x_2) \frac{d}{dx_2} \sum_{n=0}^{\infty} \frac{\Gamma(3n+1/2)}{(-9i)^n(2n)!\,[x_2\exp(-i\pi)]^{3n/2}}$$

$$+ \frac{\exp(-i\pi/4)}{4\pi} [f_2(x_2)]^2 \frac{d}{dx_2} \sum_{n=0}^{\infty} \frac{\Gamma(3n+1/2)}{(9i)^n(2n)!\,[x_2\exp(-i\pi)]^{3n/2}}$$

$$= \frac{\exp(-i\pi/4)}{2\pi} \sum_{n=0}^{\infty} \frac{\Gamma(3n+1/2)}{(2n)! \left[\dfrac{27i}{2}\zeta(x_2) \right]^n}$$

$$+ \frac{27\exp(+3i\pi/4)}{16\pi} \sum_{n=1}^{\infty} \frac{\Gamma(3n+1/2)}{(2n-1)! \left[\dfrac{27i}{2}\zeta(x_2) \right]^{n+1}}$$

$$+ \frac{27\exp(+i\pi/4)}{16\pi} \exp[-2i\zeta(x_2)] \sum_{n=1}^{\infty} \frac{\Gamma(3n+1/2)}{(2n-1)! \left[-\dfrac{27i}{2}\zeta(x_2) \right]^{n+1}}$$

$$= \frac{\exp(-i\pi/4)}{2\pi^{1/2}} \left[1 + \frac{5i}{72\,|\zeta(x_2)|} + \cdots \right], \qquad (3.17.16a)$$

$$\bar{a}_2(x_2) = \bar{a}_1^*(x_2) \sim \frac{\exp(+i\pi/4)}{2\pi^{1/2}} \left[1 - \frac{5i}{72\,|\zeta(x_2)|} + \cdots \right]. \qquad (3.17.16b)$$

3.18 Expressions for the parameters α, β, and γ when $Q^2(z) = R(z) = -z$

The parameters α, β and γ, defined in Section 2.4, appear in our discussion of the connection formulas, and it is therefore of interest to derive approximate expressions for these parameters for a particular case with general applicability. Calculate therefore, in the first-order approximation with $Q^2(z) = R(z)$, approximate expressions for the parameters α, β and γ pertaining to the Airy differential equation (3.17.1).

Solution. In Section 3.17 we obtained expressions for the a-coefficients in the classically forbidden region as well as in the classically allowed region for the particular case that $Q^2(z) = R(z) = -z$. Using these expressions we start by obtaining two elements of the matrix $\mathbf{F}(x_1, x_2)$ connecting a point x_2 in the classically allowed region and a point x_1 in the classically forbidden region along a path in the upper half of the complex z-plane, proceeding far enough away from the turning point t in order that the μ-integral $\mu(x_1, x_2)$ be small compared with unity.

Noting that

$$\begin{vmatrix} \text{Ai}(z) & \text{Bi}(z) \\ \text{Ai}'(z) & \text{Bi}'(z) \end{vmatrix} = \frac{1}{\pi}, \qquad (3.18.1)$$

we obtain from (2.3.35) with $\psi(z) = \text{Ai}(z)$ and $\bar{\psi}(z) = \text{Bi}(z)$ the formula (cf. (3.17.2a,b) and (3.17.3a,b))

$$\mathbf{F}(x_1, x_2)$$
$$= -2\pi i \begin{pmatrix} a_1(x_1)\bar{a}_2(x_2) - \bar{a}_1(x_1)a_2(x_2) & -a_1(x_1)\bar{a}_1(x_2) + \bar{a}_1(x_1)a_1(x_2) \\ a_2(x_1)\bar{a}_2(x_2) - \bar{a}_2(x_1)a_2(x_2) & -a_2(x_1)\bar{a}_1(x_2) + \bar{a}_2(x_1)a_1(x_2) \end{pmatrix}.$$

$$(3.18.2)$$

From (3.18.2) and using (3.17.13a,b), (3.17.14a,b), (3.17.15b) and (3.17.16b) we obtain

$$F_{11}(x_1, x_2) \approx 1 + \frac{5}{72\,|\zeta(x_1)|} - \frac{5i}{72\,|\zeta(x_2)|}, \qquad (3.18.3a)$$

$$F_{21}(x_1, x_2) \approx \frac{5\exp[2\,|\zeta(x_1)|]}{144\,|\zeta(x_1)|^2}. \qquad (3.18.3b)$$

With the aid of the symmetry relations (2.4.2a,b) one can obtain from (3.18.3a,b) expressions for $F_{12}(x_1, x_2)$ and $F_{22}(x_1, x_2)$, but they are not needed in the present context.

Recalling Fig. 2.4.2 and the expressions (2.4.7a,b,c'') for α, β and γ, we obtain with the aid of (3.18.3a,b)

$$\alpha = |F_{11}(x_1, x_2)| \approx 1 + \frac{5}{72\,|\zeta(x_1)|}, \qquad (3.18.4a)$$

$$\beta = -\arg F_{11}(x_1, x_2) \approx \frac{5}{72\,|\zeta(x_2)|}, \qquad (3.18.4b)$$

$$\gamma = \text{Re}\,\frac{F_{21}(x_1, x_2)}{F_{11}(x_1, x_2)} \approx \frac{5\exp[2|\zeta(x_1)|]}{144|\zeta(x_1)|^2}. \qquad (3.18.4c)$$

The parameter γ associated with the Airy differential equation is thus always large compared with unity when $|\zeta(x_1)| \gg 1$. For other differential equations γ may change sign and hence be equal to zero at an exceptional point.

If $x_2 = -x_1$ we have, according to (3.17.7b) and (3.17.8b), $|\zeta(x_1)| = |\zeta(x_2)| = |\zeta|$ and, according to (2.2.1), with $Q^2(z) = R(z) = -z$, $|\varepsilon_0| = 5/(36\,|\zeta|^2)$, and the first-order μ-integral (2.3.18) along a half-circle with its centre at $z = 0$, which connects x_1 and x_2, is

$$\mu = |\varepsilon_0| \times |Q| \times \pi |z| = \frac{5\pi |z|^{3/2}}{36|\zeta|^2} = \frac{5\pi}{24|\zeta|}. \qquad (3.18.5)$$

Therefore we can write (3.18.4a–c) as

$$\alpha \approx 1 + \mu/(3\pi) \approx 1 + 0.1\mu, \qquad (3.18.6a)$$

$$\beta \approx \mu/(3\pi) \approx 0.1\mu, \qquad (3.18.6b)$$

$$\gamma = \frac{4\mu^2}{5\pi^2} \exp\left[2\,|\zeta(x_1)|\right] \approx 0.1\mu^2 \exp[2|\zeta(x_1)|]. \qquad (3.18.6c)$$

For the Airy differential equation the estimates (2.4.8a,b) are thus rather good, while the upper limit (2.4.8c) for γ is considerably larger than the result (3.18.6c). However, the approximate value of γ obtained from (3.18.6c) is also large compared with unity.

3.19 Solutions of the Airy differential equation that at a fixed point on one side of the turning point are represented by a single, *pure* phase-integral function, and their representation on the other side of the turning point

The purpose of the treatment in this problem is to derive formulas that will be used in the discussion in Section 3.20 of the connection formulas accociated with the turning point of the Airy differential equation in order to illuminate in a concrete and indisputable way their one-directional nature. Consider therefore a solution of the Airy differential equation $d^2\psi/dz^2 - z\psi = 0$ that at a given, *fixed* point on one side of the turning point is represented by a single, *pure* phase-integral function ($f_1(z)$ or $f_2(z)$), without any admixture of the other phase-integral function. Find the phase-integral expression for this wave function on the other side of the turning point in the following cases:

(a) The given point lies in the classically forbidden region, and the solution at this point is represented by the dominant phase-integral function.
(b) The given point lies in the classically forbidden region, and the solution at this point is represented by the subdominant phase-integral function.
(c) The given point lies in the classically allowed region, and the solution at this point is represented by the phase-integral function corresponding to a wave travelling away from the turning point.
(d) The given point lies in the classically allowed region, and the solution at this point is represented by the phase-integral function corresponding to a wave travelling towards the turning point.

Solution. We recall (3.17.2a,b) and (3.17.3a,b), which can be summarized as

$$\begin{pmatrix} \text{Ai}(z) & \text{Bi}(z) \\ \text{Ai}'(z) & \text{Bi}'(z) \end{pmatrix} = \begin{pmatrix} f_1(z) & f_2(z) \\ f_1'(z) & f_2'(z) \end{pmatrix} \begin{pmatrix} a_1(z) & \bar{a}_1(z) \\ a_2(z) & \bar{a}_2(z) \end{pmatrix}. \tag{3.19.1}$$

From (3.19.1), (3.18.1) and (2.3.2) it follows that the determinant of the a-coefficient matrix in (3.19.1) is equal to $-1/(2\pi i)$ and hence

$$\begin{pmatrix} a_1(z) & \bar{a}_1(z) \\ a_2(z) & \bar{a}_2(z) \end{pmatrix}^{-1} = -2\pi i \begin{pmatrix} \bar{a}_2(z) & -\bar{a}_1(z) \\ -a_2(z) & a_1(z) \end{pmatrix}. \tag{3.19.2}$$

From (3.19.1) and (3.19.2) it follows that

$$\begin{pmatrix} f_1(z) & f_2(z) \\ f_1'(z) & f_2'(z) \end{pmatrix} = -2\pi i \begin{pmatrix} \text{Ai}(z) & \text{Bi}(z) \\ \text{Ai}'(z) & \text{Bi}'(z) \end{pmatrix} \begin{pmatrix} \bar{a}_2(z) & -\bar{a}_1(z) \\ -a_2(z) & a_1(z) \end{pmatrix}, \tag{3.19.3}$$

i.e.,

$$f_1(z) = -2\pi i \bar{a}_2(z)\text{Ai}(z) + 2\pi i a_2(z)\text{Bi}(z), \tag{3.19.4a}$$

$$f_1'(z) = -2\pi i \bar{a}_2(z)\text{Ai}'(z) + 2\pi i a_2(z)\text{Bi}'(z), \tag{3.19.4b}$$

$$f_2(z) = 2\pi i \bar{a}_1(z)\text{Ai}(z) - 2\pi i a_1(z)\text{Bi}(z), \tag{3.19.5a}$$

$$f_2'(z) = 2\pi i \bar{a}_1(z)\text{Ai}'(z) - 2\pi i a_1(z)\text{Bi}'(z). \tag{3.19.5b}$$

As in Section 3.17 we choose

$$Q^2(z) = R(z) = -z = |z| \exp[i(\arg z - \pi)], \quad 0 \le \arg z \le \pi, \tag{3.19.6}$$

and consider the first-order approximation. Letting $x_1 (>0)$ be a point in the classically forbidden region and $x_2 (<0)$ be a point in the classically allowed region, we have according to (3.17.4a,b), (3.17.7a,b) and (3.17.8a,b) the following formulas for the phase-integral functions:

$$\left| Q^{-1/2}(x_1) \right| \exp[+|\zeta(x_1)|] = \exp(-i\pi/4) f_1(x_1), \tag{3.19.7a}$$

$$\left| Q^{-1/2}(x_1) \right| \exp[-|\zeta(x_1)|] = \exp(-i\pi/4) f_2(x_1), \tag{3.19.7b}$$

$$\left| Q^{-1/2}(x_2) \right| \exp[+i|\zeta(x_2)|] = f_2(x_2), \tag{3.19.8a}$$

$$\left| Q^{-1/2}(x_2) \right| \exp[-i|\zeta(x_2)|] = f_1(x_2). \tag{3.19.8b}$$

Case (a)

The solution $\psi(z)$ that at the *fixed* point x_1 in the classically forbidden region is represented by the dominant phase-integral function, without any admixture of the subdominant phase-integral function, i.e., the solution $\psi(z)$ that satisfies the initial conditions (cf. (3.19.7a))

$$\psi(x_1) = \left|Q^{-1/2}(x_1)\right|\exp[+|\zeta(x_1)|] = \exp(-i\pi/4)f_1(x_1), \quad (3.19.9a)$$

$$\psi'(x_1) = \frac{d}{dx_1}\{\left|Q^{-1/2}(x_1)\right|\exp[+|\zeta(x_1)|]\} = \exp(-i\pi/4)f_1'(x_1), \quad (3.19.9b)$$

is according to (3.19.4a,b) at an arbitrary point z given by

$$\psi(z) = \exp(-i\pi/4)\left[-2\pi i\bar{a}_2(x_1)\mathrm{Ai}(z) + 2\pi i a_2(x_1)\mathrm{Bi}(z)\right]. \quad (3.19.10)$$

Inserting into (3.19.10) the asymptotic expressions (3.17.13b) and (3.17.14b) for $a_2(x_1)$ and $\bar{a}_2(x_1)$, respectively, we obtain

$$\psi(z) \approx -\frac{5\pi^{1/2}\exp[2\,|\zeta(x_1)|]}{72\,|\zeta(x_1)|^2}\mathrm{Ai}(z) + \pi^{1/2}\,\mathrm{Bi}(z), \quad (3.19.11)$$

the coefficient of $\mathrm{Bi}(z)$ being negligible compared with the error of the coefficient of $\mathrm{Ai}(z)$. Proceeding by inserting into (3.19.11), with z chosen to be a point x_2 in the classically allowed region, the approximate expressions (3.17.12a',b') for $\mathrm{Ai}(x_2)$ and $\mathrm{Bi}(x_2)$, we obtain

$$\psi(x_2) \approx \frac{5\exp[2\,|\zeta(x_1)|]}{72|\zeta(x_1)|^2}\left|Q^{-1/2}(x_2)\right|\cos[|\zeta(x_2)| + 3\pi/4]. \quad (3.19.12)$$

In a paper concerning Borel summation of JWKB expansions of solutions of the Airy differential equation, Silverstone (1985) seemed to suppose that in the classically forbidden region $\sqrt{\pi}\,\mathrm{Bi}(z)$ is represented by the *pure* dominant phase-integral function. It is, however, seen from (3.19.11) that the solution $\psi(z)$ that at the point x_1 in the classically forbidden region is represented by the *pure* phase-integral function $|Q^{-1/2}(z)|\exp[+|\zeta(z)|]$ is *not* $\sqrt{\pi}\,\mathrm{Bi}(z)$. On the contrary there is a *dominant* admixture in $\psi(z)$ of the function $\mathrm{Ai}(z)$, and this admixture depends strongly on the position of the point x_1.

Case (b)

The solution $\psi(z)$ that at the *fixed* point x_1 in the classically forbidden region is represented by the subdominant phase-integral function, without any admixture of the dominant phase-integral function, i.e., the solution $\psi(z)$ that fulfils the initial conditions (cf. (3.19.7b))

$$\psi(x_1) = \left|Q^{-1/2}(x_1)\right|\exp[-|\zeta(x_1)|] = \exp(-i\pi/4)f_2(x_1), \quad (3.19.13a)$$

$$\psi'(x_1) = \frac{d}{dx_1}\{\left|Q^{-1/2}(x_1)\right|\exp[-|\zeta(x_1)|]\} = \exp(-i\pi/4)f_2'(x_1), \quad (3.19.13b)$$

is, according to (3.19.5a,b), given at an arbitrary point z by

$$\psi(z) = \exp(-i\pi/4)[2\pi i\bar{a}_1(x_1)\mathrm{Ai}(z) - 2\pi i a_1(x_1)\mathrm{Bi}(z)]. \quad (3.19.14)$$

Proceeding in the same way as in Case (a), we insert (3.17.13a) and (3.17.14a) into (3.19.14), and obtain

$$\psi(z) \approx 2\pi^{1/2} \text{Ai}(z) - \frac{5\pi^{1/2} \exp[-2|\zeta(x_1)|]}{144|\zeta(x_1)|^2} \text{Bi}(z), \qquad (3.19.15)$$

the coefficient of Bi(z) being negligible compared with the error of the coefficient of Ai(z). Choosing z to be a point x_2 in the classically allowed region, and using the asymptotic approximations (3.17.12a',b') for Ai(x_2) and Bi(x_2), we obtain from (3.19.15)

$$\psi(x_2) \approx 2|Q^{-1/2}(x_2)| \cos[|\zeta(x_2)| - \pi/4]. \qquad (3.19.16)$$

Case (c)

The solution $\psi(z)$ that at the *fixed* point x_2 in the classically allowed region is represented by the phase-integral function corresponding to a wave travelling away from the turning point, without any admixture of the phase-integral function that corresponds to a wave travelling towards the turning point, i.e., the solution $\psi(z)$ that satisfies the initial conditions (cf. (3.19.8a))

$$\psi(x_2) = |Q^{-1/2}(x_2)| \exp[+i|\zeta(x_2)|] = f_2(x_2), \qquad (3.19.17a)$$

$$\psi'(x_2) = \frac{d}{dx_2} \{|Q^{-1/2}(x_2)| \exp[+i|\zeta(x_2)|]\} = f_2'(x_2), \qquad (3.19.17b)$$

is, according to (3.19.5a,b), given at an arbitrary point z by

$$\psi(z) = 2\pi i \bar{a}_1(x_2)\text{Ai}(z) - 2\pi i a_1(x_2)\text{Bi}(z). \qquad (3.19.18)$$

Inserting into (3.19.18) the asymptotic expressions (3.17.15a) and (3.17.16a) for $a_1(x_2)$ and $\bar{a}_1(x_2)$, respectively, we get approximately

$$\psi(z) \approx \pi^{1/2} \left[1 + \frac{5i}{72|\zeta(x_2)|}\right] [\exp(+i\pi/4)\text{Ai}(z) + \exp(-i\pi/4)\text{Bi}(z)]. \qquad (3.19.19)$$

When z is a point x_1 in the classically forbidden region, we obtain from (3.19.19) with the aid of the asymptotic formulas (3.17.11a',b') for Ai(x_1) and Bi(x_1)

$$\psi(x_1) \approx \left[1 + \frac{5i}{72|\zeta(x_2)|}\right] \{\tfrac{1}{2}\exp(+i\pi/4)|Q^{-1/2}(x_1)| \exp[-|\zeta(x_1)|]$$

$$+ \exp(-i\pi/4)|Q^{-1/2}(x_1)| \exp[+|\zeta(x_1)|]\}$$

$$\approx \left[1 + \frac{5i}{72|\zeta(x_2)|}\right] \exp(-i\pi/4)|Q^{-1/2}(x_1)| \exp[+|\zeta(x_1)|], \qquad (3.19.20)$$

the subdominant term being negligible compared with the error of the dominant term.

Case (d)

The solution $\psi(z)$ that at the *fixed* point x_2 in the classically allowed region is represented by the phase-integral function corresponding to a wave travelling towards the turning point, without any admixture of the phase-integral function that corresponds to a wave travelling away from the turning point satisfies initial conditions that are complex conjugates of (3.19.17a,b), i.e., since the complex conjugate of $f_2(x_2)$ is $f_1(x_2)$ according (3.19.8a,b),

$$\psi(x_2) = \left|Q^{-1/2}(x_2)\right|\exp[-i|\zeta(x_2)|] = f_1(x_2), \qquad (3.19.21a)$$

$$\psi'(x_2) = \frac{d}{dx_2}\left\{\left|Q^{-1/2}(x_2)\right|\exp[-i|\zeta(x_2)|]\right\} = f_1'(x_2). \qquad (3.19.21b)$$

For points x_1 in the classically forbidden region this solution is given by the complex conjugate of (3.19.20) i.e., by

$$\psi(x_1) \approx \left[1 - \frac{5i}{72|\zeta(x_2)|}\right]\exp(+i\pi/4)\left|Q^{-1/2}(x_1)\right|\exp[+|\zeta(x_1)|]. \quad (3.19.22)$$

3.20 Connection formulas and their one-directional nature demonstrated for the Airy differential equation

Consider the Airy differential equation $d^2\psi/dz^2 - z\psi = 0$ and use the results obtained in Section 3.19 to derive, in the first-order approximation with $Q^2(z) = R(z) = -z$, the connection formulas and to demonstrate their one-directional nature.

Solution. The solution $\psi(z)$ that at the point $x_1\,(>0)$ in the classically forbidden region is given by (3.19.13a,b) is at the point $x_2\,(<0)$ in the classically allowed region approximately given by (3.19.16). This yields the connection formula

$$\left|Q^{-1/2}(x)\right|\exp[-|\zeta(x)|] \to 2\left|Q^{-1/2}(x)\right|\cos[|\zeta(x_2)| - \pi/4], \quad (3.20.1)$$

i.e., the connection formula (3.10.10′) in the first-order approximation, derived here for the Airy differential equation. To demonstrate its one-directional nature we proceed as follows. The solution that is given by the expression on the right-hand side of (3.20.1) at the point $x_2\,(<0)$ in the classically allowed region, i.e.,

$$\psi(x_2) = \exp(-i\pi/4)\left|Q^{-1/2}(x_2)\right|\exp[+i|\zeta(x_2)|]$$

$$+ \exp(+i\pi/4)\left|Q^{-1/2}(x_2)\right|\exp[-i|\zeta(x_2)|], \qquad (3.20.2)$$

is, according to (3.19.20) and (3.19.22), at the point x_1 (> 0) in the classically forbidden region approximately

$$\psi(x_1) = -i\left[1 + \frac{5i}{72|\zeta(x_2)|}\right]\left|Q^{-1/2}(x_1)\right|\exp[+|\zeta(x_1)|]$$

$$+ i\left[1 - \frac{5i}{72|\zeta(x_2)|}\right]\left|Q^{-1/2}(x_1)\right|\exp[+|\zeta(x_1)|]$$

$$= \frac{5}{36|\zeta(x_2)|}\left|Q^{-1/2}(x_1)\right|\exp[+|\zeta(x_1)|]. \tag{3.20.3}$$

The right-hand side of (3.20.3) differs drastically from the left-hand side of (3.20.1), and the one-directional nature of the connection formula (3.10.10′) is thus demonstrated explicitly in the first-order approximation for the Airy differential equation.

From (3.19.17a,b) and (3.19.20), neglecting $5i / [72\,|\zeta(x_2)|]$, we obtain the connection formula

$$\left|Q^{-1/2}(x)\right|\exp\{+i[|\zeta(x)| + \pi/4]\} \rightarrow \left|Q^{-1/2}(x)\right|\exp[+|\zeta(x)|], \quad (3.20.4a)$$

and from (3.19.21a,b) and (3.19.22), neglecting $5i / [72\,|\zeta(x_2)|]$, we obtain the connection formula

$$\left|Q^{-1/2}(x)\right|\exp\{-i[|\zeta(x)| + \pi/4]\} \rightarrow \left|Q^{-1/2}(x)\right|\exp[+|\zeta(x)|]. \quad (3.20.4b)$$

It may at first sight seem strange that in these connection formulas the left-hand sides are complex, while the right-hand sides are real. The explanation of this is given below the connection formulas (3.12.7), which in the first order agree with (3.20.4a,b).

The solution that at the *fixed* point x_2 (< 0) in the classically allowed region is given by $|Q^{-1/2}(x_2)| \cos[|\zeta(x_2)| + \delta - \pi/4]$, without admixture of any other phase-integral function, i.e., the solution $\psi(z)$ that satisfies the initial conditions

$$\psi(x_2) = \left|Q^{-1/2}(x_2)\right| \cos[|\zeta(x_2)| + \delta - \pi/4], \tag{3.20.5a}$$

$$\psi'(x_2) = \frac{d}{dx_2}\{|Q^{-1/2}(x_2)| \cos[|\zeta(x_2)| + \delta - \pi/4]\}, \tag{3.20.5b}$$

i.e.,

$$\psi(x_2) = \tfrac{1}{2}\exp[i(\delta - \pi/4)]\left|Q^{-1/2}(x_2)\right|\exp[i|\zeta(x_2)|]$$

$$+ \tfrac{1}{2}\exp[-i(\delta - \pi/4)]\left|Q^{-1/2}(x_2)\right|\exp[-i|\zeta(x_2)|], \quad (3.20.5a')$$

$$\psi'(x_2) = \tfrac{1}{2}\exp[i(\delta - \pi/4)]\frac{d}{dx_2}\{|Q^{-1/2}(x_2)|\exp[i|\zeta(x_2)|]\}$$

$$+ \tfrac{1}{2}\exp[-i(\delta - \pi/4)]\frac{d}{dx_2}\{|Q^{-1/2}(x_2)|\exp[-i|\zeta(x_2)|]\},$$

$$\tag{3.20.5b'}$$

is, according to (3.19.20) and (3.19.22), at the point x_1 in the classically forbidden region approximately given by the formula

$$\psi(x_1) = \left[\sin\delta + \frac{5\cos\delta}{72|\zeta(x_2)|}\right]|Q^{-1/2}(x_1)|\exp[+|\zeta(x_1)|]. \qquad (3.20.6)$$

When $|\tan\delta| \gg 5/[72|\zeta(x_2)|]$ we obtain from (3.20.6) the approximate formula

$$\psi(x_1) = \sin\delta\,|Q^{-1/2}(x_1)|\exp[+|\zeta(x_1)|], \qquad (3.20.7a)$$

which, together with (3.20.5a), can be written in the form of the connection formula

$$|Q^{-1/2}(x)|\cos[|\zeta(x)| + \delta - \pi/4] \to \sin\delta\,|Q^{-1/2}(x)|\exp[+|\zeta(x)|], \quad (3.20.8)$$

which agrees with the first order of the connection formula (3.12.9). When $|\tan\delta| \ll 5/[72|\zeta(x_2)|]$ we obtain instead of (3.20.7a) the approximate formula

$$\psi(x_1) = \frac{5\cos\delta}{72|\zeta(x_2)|}|Q^{-1/2}(x_1)|\exp[+|\zeta(x_1)|]. \qquad (3.20.7b)$$

To demonstrate the one-directional nature of the connection formula (3.20.8) we proceed as follows. According to (3.19.9a,b) and (3.19.12) the solution that is given by the expression on the right-hand side of (3.20.8) at the point $x_1 (> 0)$ in the classically forbidden region is at the point $x_2 (< 0)$ in the classically allowed region approximately

$$\psi(x_2) = \sin\delta\frac{5\exp[2|\zeta(x_1)|]}{72|\zeta(x_1)|^2}|Q^{-1/2}(x_2)|\cos[|\zeta(x_2)| + 3\pi/4]. \quad (3.20.9)$$

The amplitude on the right-hand side of (3.20.9) differs drastically from the amplitude on the left-hand side of (3.20.8), and the phase of the cosine in (3.20.9) differs by $\pi - \delta$ from the phase of the cosine in (3.20.8). The one-directional nature of the connection formula (3.20.8) is thus demonstrated for the Airy differential equation and the first order of the phase-integral approximation. When we put $C(x_1) = \sin\delta$ and $D(x_1) = 0$ in (3.10.2) and (3.10.13), comparison with (3.20.8) and (3.20.9) shows that for the Airy differential equation and in the first-order approximation the parameter γ is given by

$$\gamma = 5\exp[2|\zeta(x_1)|]/[144|\zeta(x_1)|^2] \qquad (3.20.10)$$

in agreement with (3.18.4c). One can also obtain this expression for γ by comparing (3.19.12) with the first order of (3.13.4a) with $\delta = \pi/2$.

3.21 Dependence of the phase of the wave function in a classically allowed region on the value of the logarithmic derivative of the wave function at a fixed point x_1 in an adjacent classically forbidden region

For a real, smooth potential, consider a simple, well-isolated turning point t. At a point x_1 well inside the classically forbidden region we impose the boundary condition that the logarithmic derivative of the wave function has a real value, k. Investigate how the phase of the wave function at a point x_2 in the classically allowed region changes when k is changed, while x_1 is kept fixed. This problem cannot be solved within the framework of asymptotic analysis, and its solution is therefore one of the most important applications of the F-matrix method and the parameterization of the F-matrix.

Solution. We consider simultaneously the cases in Figs. 2.4.1 and 2.4.2 with $w(z)$ defined by (2.4.1). If we impose the boundary condition that the logarithmic derivative of the wave function is equal to k at the point x_1, i.e.,

$$\psi'(x_1)/\psi(x_1) = k, \tag{3.21.1}$$

we obtain with the aid of (2.3.4a,b) the condition

$$\frac{a_1(x_1)f_1'(x_1) + a_2(x_1)f_2'(x_1)}{a_1(x_1)f_1(x_1) + a_2(x_1)f_2(x_1)} = k. \tag{3.21.2}$$

From this, with the aid of the definitions (2.3.1a,b) and (2.4.1) of $f_1(z)$ and $f_2(z)$, and with due regard to the phases in Fig. 2.4.1 and Fig. 2.4.2, we obtain

$$a_1(x_1) = -a_2(x_1)\frac{kf_2(x_1) - f_2'(x_1)}{kf_1(x_1) - f_1'(x_1)}$$

$$= -a_2(x_1)\exp[-2iw(x_1)]\frac{k + iq(x_1) + q'(x_1)/[2q(x_1)]}{k - iq(x_1) + q'(x_1)/[2q(x_1)]}$$

$$= \kappa(k)\,a_2(x_1)\exp[-2iw(x_1)] = \kappa(k)\,a_2(x_1)\exp[-2|w(x_1)|], \tag{3.21.3}$$

where

$$\kappa(k) = \frac{1 \mp k/|q(x_1)| \pm \frac{1}{2}d[1/|q(x_1)|]/dx_1}{1 \pm k/|q(x_1)| \mp \frac{1}{2}d[1/|q(x_1)|]/dx_1}$$

$$= \frac{2}{1 \pm k/|q(x_1)| \mp \frac{1}{2}d[1/|q(x_1)|]/dx_1} - 1, \tag{3.21.4}$$

the upper and the lower signs being appropriate for the cases in Fig. 2.4.1 and Fig. 2.4.2, respectively. Since k is real, the function $\kappa(k)$ is also real. It is monotonically decreasing for the upper signs (Fig. 2.4.1), but monotonically increasing for the lower signs (Fig. 2.4.2).

Using (2.3.7), (2.3.12c) and (3.21.3), we obtain

$$a_1(x_2) = F_{11}(x_2, x_1)a_1(x_1) + F_{12}(x_2, x_1)a_2(x_1)$$
$$= F_{22}(x_1, x_2)a_1(x_1) - F_{12}(x_1, x_2)a_2(x_1)$$
$$= \{\kappa(k)\exp[-2i\,w(x_1)]\,F_{22}(x_1, x_2) - F_{12}(x_1, x_2)\}a_2(x_1). \quad (3.21.5)$$

Inserting $F_{12}(x_1, x_2)$ and $F_{22}(x_1, x_2)$, expressed in terms of the parameters α, β and γ according to (2.4.9b,d), into (3.21.5), we get

$$a_1(x_2) = \alpha \exp[\mp i(\pi/2 + \beta)]\{1 - \kappa(k)\gamma\exp[-2|w(x_1)|]\}$$
$$\pm \frac{i\kappa(k)}{2\alpha^2}\exp[-2|w(x_1)|]\}a_2(x_1), \quad (3.21.6)$$

which yields

$$\arg a_1(x_2) = \arg a_2(x_1) \mp (\pi/2 + \beta)$$
$$\pm \arctan\frac{\kappa(k)\exp[-2|w(x_1)|]}{2\alpha^2\{1 - \kappa(k)\,\gamma\exp[-2|w(x_1)|]\}}, \quad (3.21.7)$$

where an appropriate branch of arctan is to be chosen.

When we choose the wave function to be real, we have $a_2(x_2) = a_1^*(x_2)$, and according to (2.3.4a) with (2.3.1a,b) we can, with due regard to the phases of $w(x_2)$ displayed in Fig. 2.4.1 and Fig. 2.4.2, write the wave function at the point x_2 in the classically allowed region as

$$\psi(x_2) = a_1(x_2)f_1(x_2) + a_2(x_2)f_2(x_2)$$
$$= q^{-1/2}(x_2)\{a_1(x_2)\exp[i\,w(x_2)] + a_1^*(x_2)\exp[-i\,w(x_2)]\}$$
$$= 2|a_1(x_2)|q^{-1/2}(x_2)\cos[w(x_2) + \arg a_1(x_2)]$$
$$= 2|a_1(x_2)|\left|q^{-1/2}(x_2)\right|\cos[\pm|w(x_2)| + \arg a_1(x_2)]$$
$$= 2|a_1(x_2)|\left|q^{-1/2}(x_2)\right|\cos[|w(x_2)| \pm \arg a_1(x_2)],$$

i.e.,

$$\psi(x_2) = \Omega(x_2)\left|q^{-1/2}(x_2)\right|\cos[|w(x_2)| + \delta(k) - \pi/4], \quad (3.21.8)$$

where

$$\Omega(x_2) = 2|a_1(x_2)|, \quad (3.21.9a)$$
$$\delta(k) = \pi/4 \pm \arg a_1(x_2). \quad (3.21.9b)$$

Using (3.21.7) we can write (3.21.9b) as

$$\delta(k) = \pm \arg a_2(x_1) - (\pi/4 + \beta) + \arctan \frac{\kappa(k)\exp[-2|w(x_1)|]}{2\alpha^2\{1 - \kappa(k)\gamma\exp[-2|w(x_1)|]\}},$$

(3.21.10)

where, as in (3.21.7), an appropriate branch of arctan is to be chosen. The term in (3.21.10) that contains $\kappa(k)$ is a monotonically increasing function of $\kappa(k)$. Recalling the monotonicity properties of $\kappa(k)$ stated below (3.21.4), we therefore realize that $\delta(k)$ is a monotonically decreasing function of k for the case in Fig. 2.4.1, but a monotonically increasing function of k for the case in Fig. 2.4.2.

To change the value of the logarithmic derivative of the wave function at the point x_1 it is sufficient to change one of the a-coefficients in (3.21.2). We can therefore assume that $a_2(x_1)$ is kept fixed. When k is changed from $\bar{k}$ to k, while $x_1, a_2(x_1)$ and x_2 are kept fixed, we obtain from (3.21.10)

$$\delta(k) - \delta(\bar{k}) = \arctan \frac{\kappa(k)\exp[-2|w(x_1)|]}{2\alpha^2\{1 - \kappa(k)\gamma\exp[-2|w(x_1)|]\}}$$

$$- \arctan \frac{\kappa(\bar{k})\exp[-2|w(x_1)|]}{2\alpha^2\{1 - \kappa(\bar{k})\gamma\exp[-2|w(x_1)|]\}}$$

$$= \arctan \frac{\left[\kappa(k) - \kappa(\bar{k})\right]e^{-2|w(x_1)|}}{2\alpha^2[1 - \kappa(k)\gamma e^{-2|w(x_1)|}][1 - \kappa(\bar{k})\gamma e^{-2|w(x_1)|}] + [\kappa(k)\kappa(\bar{k})/(2\alpha^2)]e^{-4|w(x_1)|}}.$$

(3.21.11)

Recalling (3.21.4), we obtain
$$\kappa(k) - \kappa(\bar{k})$$

$$= \frac{\mp 2\left(k - \bar{k}\right)/|q(x_1)|}{\{1 \pm k/|q(x_1)| \mp \frac{1}{2}d[1/|q(x_1)|]/dx_1\}\{1 \pm \bar{k}/|q(x_1)| \mp \frac{1}{2}d[1/|q(x_1)|]/dx_1\}},$$

(3.21.12)

where the upper and the lower signs apply to the cases in Figs. 2.4.1 and 2.4.2, respectively. Formula (3.21.11) along with (3.21.4) or (3.21.12) forms the basis for the solution of intricate physical problems; see Sections 3.22, 3.23, 3.24 and 3.29.

3.22 Phase of the wave function in the classically allowed regions adjacent to a real, *symmetric* potential barrier, when the logarithmic derivative of the wave function is given at the centre of the barrier

(a) Use the formulas in Section 2.5 to obtain the phase of the wave function in the classically allowed regions on opposite sides of a real, *symmetric*, single-hump potential barrier *of arbitrary thickness,* when either the wave function or its derivative is equal to zero at the centre of the barrier.

(b) Use the results obtained in (a) along with (3.21.11), (3.21.4) and (3.21.12) to obtain formulas for the phase of the wave function in the classically allowed regions adjacent to a real, *thick, symmetric* single-hump potential barrier, when the logarithmic derivative of the wave function at the centre of the barrier is known to be equal to k (real).

Solution. Consider a symmetric, real single-hump potential barrier with its centre at the point x_1 and assume that

$$\psi'(x_1)/\psi(x_1) = k, \quad k \text{ real.} \tag{3.22.1}$$

Instead of the notation δ' and δ'' used in Section 2.5 for the phase constants in the wave function on opposite sides of the barrier, we now write $\delta'(k)$ and $\delta''(k)$. The phase $\delta'(k)$ is only determined modulo π, since according to (3.22.1) k remains unchanged when the wave function changes sign, while the difference $\delta'(k) - \delta''(k)$ is determined modulo 2π. When k changes continuously, $\delta'(k)$ and $\delta''(k)$ also change continuously.

Case (a)
When $Q^2(z)$ is real on the real z-axis and has the same mirror symmetry as $R(z)$, and the points x_2' and x_2'' in the classically allowed regions lie mirror symmetrically with respect to the barrier, we obviously have

$$\delta'(0) = \delta''(0) \,(\text{mod } 2\pi), \tag{3.22.2a}$$

$$\delta'(\infty) = \delta''(\infty) \pm \pi \,(\text{mod } 2\pi). \tag{3.22.2b}$$

From (2.5.26) and (2.5.30) we obtain by using (2.5.8) and (3.22.2a,b)

$$\cos[2\delta'(0) - \tilde{\phi}] = \frac{\theta}{(\theta^2 + 1)^{1/2}}, \tag{3.22.3a}$$

$$\sin[2\delta'(0) - \tilde{\phi}] = \frac{1}{(\theta^2 + 1)^{1/2}}, \tag{3.22.3b}$$

$$\cos[2\delta'(\infty) - \tilde{\phi}] = \frac{\theta}{(\theta^2 + 1)^{1/2}}, \tag{3.22.4a}$$

$$\sin[2\delta'(\infty) - \tilde{\phi}] = -\frac{1}{(\theta^2 + 1)^{1/2}}, \tag{3.22.4b}$$

i.e., with the aid of (3.22.2a,b),

$$\delta'(0) = \delta''(0) = \tfrac{1}{2}\tilde{\phi} + \tfrac{1}{2}\arctan(1/\theta) \ (\text{mod } \pi), \quad 0 < \arctan < \pi/2, \tag{3.22.5}$$

$$\delta'(\infty) = \delta''(\infty) \pm \pi = \tfrac{1}{2}\tilde{\phi} - \tfrac{1}{2}\arctan(1/\theta) \ (\text{mod } \pi), \quad 0 < \arctan < \pi/2. \tag{3.22.6}$$

These formulas are valid both for superdense barriers of arbitrary thickness and for underdense barriers, since the formulas that have been used in the derivation are valid for such barriers.

When for a real, smooth, *symmetric* single-hump potential barrier of arbitrary thickness one requires that either the wave function or its derivative be equal to zero at the centre of the barrier, one obtains, according to (3.22.6) or (3.22.5), respectively, the phase of the wave function in the classically allowed regions on opposite sides of the barrier. Neglecting $\tilde{\phi}$ and $1/\theta \approx \exp(-K)$ in (3.22.5) and (3.22.6), we get $\delta'(0) = \delta''(0) \approx 0$ and $\delta'(\infty) = \delta''(\infty) \pm \pi \approx 0$, which is consistent with the result that would be obtained by using the connection formula (3.10.10), when the barrier is thick.

Case (b)

Consider now the more general case in which the logarithmic derivative k in (3.22.1) takes an arbitrary real value at the centre of the barrier. Since we assume that $Q^2(z)$ has the same symmetry as $R(z)$, ε_0 is, according (2.2.1'), also symmetric. If the barrier is superdense, and if there is no cut that crosses the real axis in the classically forbidden region, the functions $d^v \varepsilon_0/d\zeta^v = [Q^{-1}d/dz]^v \varepsilon_0$, $v = 1, 2, \ldots,$ are in that region symmetric or anti-symmetric with respect to the centre of the barrier depending on whether v is even or odd. From the structure of the functions Y_{2n} (Skorupski 1980, p. 167) it therefore follows that Y_{2n}, $n = 1, 2, \ldots,$ are symmetric in the classically forbidden region. This is thus also the case for the function $q(z)$. At the centre x_1 of the barrier we therefore have $q'(x_1) = 0$. Since, furthermore, $|w(x_1)| = K/2$, we obtain from (3.21.11) along with (3.21.4) and (3.21.12), and with due regard to Figs. 2.4.1 and 2.4.2, the change of the phase constants, when the logarithmic derivative of the wave function at the centre of the barrier changes from zero to k, as

$$\delta'(k) - \delta'(0) = -\arctan \frac{b_0' k/|q(x_1)|}{b_1' k/|q(x_1)| - b_2'}, \tag{3.22.7a}$$

$$\delta''(k) - \delta''(0) = -\arctan \frac{b_0'' k/|q(x_1)|}{b_1'' k/|q(x_1)| + b_2''}. \tag{3.22.7b}$$

One obtains b_0', b_1', b_2' and b_0'', b_1'', b_2'' in (3.22.7a,b) by replacing α, β, γ by α', β', γ' and α'', β'', γ'', respectively, in the expressions

$$b_0 = \frac{\exp(-K)}{\alpha^2[1 - \gamma \exp(-K)]} \approx \exp(-K), \tag{3.22.8a}$$

$$b_1 = 1 + \gamma \exp(-K) - \frac{\exp(-2K)}{4\alpha^4[1 - \gamma \exp(-K)]} \approx 1 + \gamma \exp(-K), \quad (3.22.8b)$$

$$b_2 = 1 - \gamma \exp(-K) + \frac{\exp(-2K)}{4\alpha^4[1 - \gamma \exp(-K)]} \approx 1 - \gamma \exp(-K). \quad (3.22.8c)$$

We have obtained the approximate expressions in (3.22.8a–c) by using (2.4.8a,c) and noting that $|w(x_1)| = K/2$ and that the barrier is assumed to be thick $[\exp(K) \gg 1]$. In an addendum to this section it is shown that $\alpha' = \alpha''$, $\beta' = \beta''$ and $\gamma' = \gamma''$. Denoting these quantities by α, β, and γ, respectively, we can thus use the formulas (3.22.8a–c) unchanged.

Since according to (2.5.10a) θ is approximately equal to $\exp(K)$, we obtain from (3.22.5) and (3.22.7a,b), on deleting (mod 2π) and the restriction on arctan,

$$\delta'(k) = \tfrac{1}{2}\tilde{\phi} + \tfrac{1}{2} \arctan \exp(-K) - \arctan \frac{b_0 k/|q(x_1)|}{b_1 k/|q(x_1)| - b_2}, \quad (3.22.9a)$$

$$\delta''(k) = \tfrac{1}{2}\tilde{\phi} + \tfrac{1}{2} \arctan \exp(-K) - \arctan \frac{b_0 k/|q(x_1)|}{b_1 k/|q(x_1)| + b_2}. \quad (3.22.9b)$$

The phase $\delta'(k)$ is a monotonically increasing function of k, while the phase $\delta''(k)$ is a monotonically decreasing function of k. From (3.22.9a,b) along with (3.22.8a–c) we obtain the approximate formulas

$$\delta'(k) = \tfrac{1}{2}\tilde{\phi} + \tfrac{1}{2} \arctan \exp(-K) - \arctan \frac{k/|q(x_1)|}{[\exp(K) + \gamma]k/|q(x_1)| - [\exp(K) - \gamma]},$$
$$(3.22.10a)$$

$$\delta''(k) = \tfrac{1}{2}\tilde{\phi} + \tfrac{1}{2} \arctan \exp(-K) - \arctan \frac{k/|q(x_1)|}{[\exp(K) + \gamma]k/|q(x_1)| + [\exp(K) - \gamma]},$$
$$(3.22.10b)$$

which for $k = \infty$ are in approximate agreement with (3.22.6); we note that the branch of arctan must not be the same in (3.22.10a) as in (3.22.10b). Letting k increase from $-\infty$ to $+\infty$, we obtain from (3.22.10a,b) the approximate values in Table 3.22.1, which is to some extent analogous to Table 2.5.1. Since according to (2.4.8c) with $|w(x_1)| = K/2$ the quantity $\gamma \exp(-K)$ is small compared with unity, the quantity $[1 - \gamma \exp(-K)]/[1 + \gamma \exp(-K)]$ that appears in Table 3.22.1 is close to unity.

Table 3.22.1. *Dependence of the phases of the wave function on opposite sides of a thick, symmetric potential barrier on the value k of the logarithmic derivative of the wave function at the centre of the barrier.*

$k/|q(x_1)|$	$\delta'(k)$	$\delta''(k)$
$-\infty$	$\frac{1}{2}\tilde{\phi} - \frac{1}{2}\exp(-K)$	$\pi + \frac{1}{2}\tilde{\phi} - \frac{1}{2}\exp(-K)$
$-\dfrac{1-\gamma\exp(-K)}{1+\gamma\exp(-K)} - \exp(-K)$	$\frac{1}{2}\tilde{\phi}$	$3\pi/4 + \frac{1}{2}\tilde{\phi} + \left(\gamma + \frac{1}{2}\right)\exp(-K)$
$-\dfrac{1-\gamma\exp(-K)}{1+\gamma\exp(-K)}$	$\frac{1}{2}\tilde{\phi}$	$\pi/2 + \frac{1}{2}\tilde{\phi} + \frac{1}{2}\exp(-k)$
$-\dfrac{1-\gamma\exp(-K)}{1+\gamma\exp(-K)} + \exp(-K)$	$\frac{1}{2}\tilde{\phi}$	$\pi/4 + \frac{1}{2}\tilde{\phi} - \left(\gamma - \frac{1}{2}\right)\exp(-k)$
0	$\frac{1}{2}\tilde{\phi} + \frac{1}{2}\exp(-K)$	$\frac{1}{2}\tilde{\phi} + \frac{1}{2}\exp(-K)$
$\dfrac{1-\gamma\exp(-K)}{1+\gamma\exp(-K)} - \exp(-K)$	$\pi/4 + \frac{1}{2}\tilde{\phi} - \left(\gamma - \frac{1}{2}\right)\exp(-K)$	$\frac{1}{2}\tilde{\phi}$
$\dfrac{1-\gamma\exp(-K)}{1+\gamma\exp(-K)}$	$\pi/2 + \frac{1}{2}\tilde{\phi} + \frac{1}{2}\exp(-K)$	$\frac{1}{2}\tilde{\phi}$
$\dfrac{1-\gamma\exp(-K)}{1+\gamma\exp(-K)} + \exp(-K)$	$3\pi/4 + \frac{1}{2}\tilde{\phi} + \left(\gamma + \frac{1}{2}\right)\exp(-K)$	$\frac{1}{2}\tilde{\phi}$
$+\infty$	$\pi + \frac{1}{2}\tilde{\phi} - \frac{1}{2}\exp(-K)$	$\frac{1}{2}\tilde{\phi} - \frac{1}{2}\exp(-K)$

Using the first-order approximation, and recalling that the barrier is assumed to be thick, we obtain from (2.5.14), (2.5.13b) and (2.5.12a) the approximate formula

$$\tilde{\phi} = \frac{1}{24\tilde{K}_0} \approx \frac{\pi}{24K}. \tag{3.22.11}$$

This quantity, as well as its error, is large compared with $\exp(-K)$. If expression (3.22.11) is used for $\tilde{\phi}$ in Table 3.22.1, the terms containing $\exp(-K)$ are therefore insignificant.

Addendum in which it is shown that the formulas obtained for $\delta'(k)$ and $\delta''(k)$ are consistent with formula (2.5.23a'), which gives the relation between the phases δ' and δ'' for a real barrier.

Since

$$\tfrac{1}{2} \arctan \exp(-K) = \arctan c \tag{3.22.12}$$

where

$$c = [\exp(2K) + 1]^{1/2} - \exp(K) = \frac{1}{[\exp(2K) + 1]^{1/2} + \exp(K)} \approx \frac{1}{2}\exp(-K),$$
$$\tag{3.22.13}$$

we can write (3.22.9a,b) as

$$\delta'(k) = \tfrac{1}{2}\tilde{\phi} + \arctan \frac{c[(b_1 - b_0/c)k/|q(x_1)| - b_2]}{(b_1 + b_0c)k/|q(x_1)| - b_2}, \tag{3.22.14a}$$

$$\delta''(k) = \tfrac{1}{2}\tilde{\phi} + \arctan \frac{c[(b_1 - b_0/c)k/|q(x_1)| + b_2]}{(b_1 + b_0c)k/|q(x_1)| + b_2}. \tag{3.22.14b}$$

From (3.22.8a,b) and (3.22.13) we obtain

$$b_1 + b_0c \approx 1 + \gamma \exp(-K) \approx 1, \tag{3.22.15a}$$
$$b_1 - b_0/c \approx -1. \tag{3.22.15b}$$

Using these formulas and (3.22.8c), we can write (3.22.14a,b) in the more explicit approximate forms

$$\delta'(k) = \tfrac{1}{2}\tilde{\phi} - \arctan \frac{c[k/|q(x_1)| + 1]}{[1 + \gamma \exp(-K)]k/|q(x_1)| - [1 - \gamma \exp(-K)]},$$
$$\tag{3.22.16a}$$

$$\delta''(k) = \tfrac{1}{2}\tilde{\phi} - \arctan \frac{c[k/|q(x_1)| - 1]}{[1 + \gamma \exp(-K)]k/|q(x_1)| + [1 - \gamma \exp(-K)]}.$$
$$\tag{3.22.16b}$$

From (3.22.16a,b) we obtain

$$\tan[\delta'(k) - \tfrac{1}{2}\tilde{\phi}]\tan[\delta''(k) - \tfrac{1}{2}\tilde{\phi}]$$

$$= \frac{c^2[k^2/|q(x_1)|^2 - 1]}{[1 + \gamma\exp(-K)]^2 k^2/|q(x_1)|^2 - [1 - \gamma\exp(-K)]^2}. \qquad (3.22.17)$$

The right-hand side of (3.22.17) is approximately equal to c^2, unless k assumes values very close to $\pm[1 - \gamma\exp(-K)]|q(x_1)|/[1 + \gamma\exp(-K)]$. Disregarding such values of k, and using (3.22.13), we therefore obtain

$$\tan\left[\delta'(k) - \tfrac{1}{2}\tilde{\phi}\right]\tan\left[\delta''(k) - \tfrac{1}{2}\tilde{\phi}\right] \approx \frac{[\exp(2K) + 1]^{1/2} - \exp(K)}{[\exp(2K) + 1]^{1/2} + \exp(K)}. \qquad (3.22.18)$$

Thus, expressions (3.22.16a,b) for the phases $\delta'(k)$ and $\delta''(k)$ approximately satisfy (2.5.23a') with θ replaced by $\exp(K)$ according to (2.5.10a) and ϑ given by (2.5.8). Hence one can approximately obtain (3.22.16a) from (2.5.23a') and (3.22.16b), and one can approximately obtain (3.22.16b) from (2.5.23a') and (3.22.16a). It is thus consistent with the discussion in Section 3.6 that (3.22.16a,b) are approximately valid independently of the sign of k. However, k must not assume values too close to $\pm[1 - \gamma\exp(-K)]\,|q(x_1)|\,/[1 + \gamma\exp(-K)]$.

Addendum concerning the parameters α, β and γ. Now we wish to find how α', β', γ' are related to α'', β'', γ''. For this purpose we note that if $\psi(x_1)$ is given by (2.4.23), $\psi(x_2)$ is given by (2.4.26), where according to (2.4.28a,b)

$$A(x_2) = \left(\frac{1}{2\alpha} + i\alpha\gamma\right)\exp[+i(\pi/4 - \beta)]C(x_1) - i\alpha\exp[+i(\pi/4 - \beta)]D(x_1),$$

$$(3.22.19a)$$

$$B(x_2) = \left(\frac{1}{2\alpha} - i\alpha\gamma\right)\exp[-i(\pi/4 - \beta)]C(x_1) + i\alpha\exp[-i(\pi/4 - \beta)]D(x_1).$$

$$(3.22.19b)$$

These formulas apply to the situation in which the turning point is t' and α, β, γ are replaced by α', β', γ', as well as to the situation in which the turning point is t'' and α, β, γ are replaced by α'', β'', γ''. If $x_2' < t'$ and $x_2'' > t''$, where $x_2' - t' = -(x_2'' - t'')$, and the coefficients $C(x_1)$ and $D(x_1)$ are the same in both situations, the relations $A(x_2') = A(x_2'')$ and $B(x_2') = B(x_2'')$ must be fulfilled for any choice of $C(x_1)$ and $D(x_1)$, since the barrier is symmetric and x_1 is situated at its centre. Using (3.22.19a,b), we therefore obtain

$$\left(\frac{1}{2\alpha'} + i\alpha'\gamma'\right)\exp[+i(\pi/4 - \beta')] = \left(\frac{1}{2\alpha''} + i\alpha''\gamma''\right)\exp[+i(\pi/4 - \beta'')],$$

$$(3.22.20a)$$

$$-i\alpha'\exp[+i(\pi/4 - \beta')] = -i\alpha''\exp[+i(\pi/4 - \beta'')], \qquad (3.22.20b)$$

and the complex conjugate relations, which, however, are consequences of (3.22.20a,b), since α', β', γ' and α'', β'', γ'' are real. We can write (3.22.20a,b) as

$$\frac{1}{2\alpha'^2} + i\gamma' = \frac{1}{2\alpha''^2} + i\gamma'', \tag{3.22.21a}$$

$$\alpha'/\alpha'' = \exp[i(\beta' - \beta'')]. \tag{3.22.21b}$$

Since α', β', γ' and α'', β'', γ'' are all real, and α', α'' are both positive, and β', β'' are small compared with unity, we obtain from (3.22.21a,b)

$$\alpha' = \alpha'', \tag{3.22.22a}$$

$$\beta' = \beta'', \tag{3.22.22b}$$

$$\gamma' = \gamma'', \tag{3.22.22c}$$

and we can thus replace α', β', γ' and α'', β'', γ'' by α, β, γ.

Alternatively (3.22.22a–c) can be derived in another way. Using (2.3.35), (2.3.36a,b) and (2.3.37a,b) along with (2.3.1a,b) and (2.4.1), and noting that $q'(x_1) = 0$, one obtains after rather lengthy calculations

$$F_{11}(x_1, x_2') = -iF_{12}(x_1, x_2''), \tag{3.22.23a}$$

$$F_{12}(x_1, x_2') = -iF_{11}(x_1, x_2''), \tag{3.22.23b}$$

$$F_{21}(x_1, x_2') = -iF_{22}(x_1, x_2''), \tag{3.22.23c}$$

$$F_{22}(x_1, x_2') = -iF_{21}(x_1, x_2''), \tag{3.22.23d}$$

where x_1 is the point situated at the centre of the symmetric barrier, and x_2' ($<t' <t''$) and x_2'' ($>t''$) are points lying symmetrically with respect to x_1. From (3.22.23a–d) one obtains the relation $\det \mathbf{F}(x_1, x_2') = \det \mathbf{F}(x_1, x_2'')$, which is consistent with (2.3.12a). Using (2.4.7a,b) along with Fig. 2.4.2, (3.22.23a), (2.4.2a) and (2.4.7a,b) along with Fig. 2.4.1 we obtain

$$\alpha'\exp(-i\beta') = F_{11}(x_1, x_2') = -iF_{12}(x_1, x_2'') = F_{11}^*(x_1, x_2'')$$
$$= [\alpha''\exp(+i\beta'')]^* = \alpha''\exp(-i\beta''), \tag{3.22.24}$$

and from this formula we obtain (3.22.22a,b), since α', α'' are positive, and β', β'' are real and small compared with unity. Using (2.4.7c''), (3.22.23a,c), (2.4.2a,b) along with Fig. 2.4.1, and again (2.4.7c'') we obtain

$$\gamma' = \mathrm{Re}\,\frac{F_{21}(x_1, x_2')}{F_{11}(x_1, x_2')} = \mathrm{Re}\,\frac{-iF_{22}(x_1, x_2'')}{-iF_{12}(x_1, x_2'')} = \mathrm{Re}\,\frac{F_{22}(x_1, x_2'')}{F_{12}(x_1, x_2'')}$$

$$= \mathrm{Re}\,\frac{iF_{21}^*(x_1, x_2'')}{iF_{11}^*(x_1, x_2'')} = \mathrm{Re}\,\left(\frac{F_{21}(x_1, x_2'')}{F_{11}(x_1, x_2'')}\right)^* = \gamma'', \tag{3.22.25}$$

i.e., (3.22.22c).

3.23 Eigenvalue problem for a quantal particle in a broad, *symmetric* potential well between two *symmetric* potential barriers of *equal shape*, with boundary conditions imposed in the *middle* of each barrier

The symmetry and the equal shape of the barriers imply that they are adjacent barriers of a truncated periodic potential. The boundary conditions to be imposed on the wave function are $\psi'(x_1)/\psi(x_1) = k$ and $\psi'(\bar{x}_1)/\psi(\bar{x}_1) = \bar{k}$, where x_1 and $\bar{x}_1(> x_1)$ denote the positions of the tops of the two barriers, respectively. Derive the quantization condition with the aid of the results obtained in Section 3.22.

Boundary conditions of the kind in question have to be imposed for the bound states of a quantal particle in a potential that is constant to the left of x_1 and to the right of $\bar{x}_1$, possibly with discontinuities at x_1 and $\bar{x}_1$. Such boundary conditions with $k = 0$ or $k = \infty$ and $\bar{k} = 0$ or $\bar{k} = \infty$ also appear in connection with the scattering of waves by an elliptic cone and related problems; see Kraus and L. M. Levine (1961) and Abawi, Dashen and H. Levine (1997). For these particular values of k and $\bar{k}$ the formulas for δ' and δ'' derived in Section 3.22 are also valid for thin barriers, and therefore the matching of WKB solutions to asymptotic solutions of a Weber differential equation, used in the treatment of such problems when the barriers are thin, is not needed.

Solution. If the turning points delimiting the broad potential well are t'' and $\bar{t}' (> t'')$, the wave function with the logarithmic derivative equal to k at the centre of the left-hand barrier can in the potential well be written as

$$\psi(x) = \Omega |q^{-1/2}(x)| \cos \left[\left| \int_{(t'')}^{x} q(z) dz \right| + \delta''(k) - \pi/4 \right], \quad \Omega > 0, \quad t'' < x < \bar{t}',$$

$$(3.23.1a)$$

where $\delta''(k)$ is given by (3.22.10b). Similarly the wave function with the logarithmic derivative equal to $\bar{k}$ at the centre of the right-hand barrier can in the potential well be written as

$$\psi(x) = \Omega |q^{-1/2}(x)| \cos \left[\left| \int_{(\bar{t}')}^{x} q(z) dz \right| + \delta'(\bar{k}) - \pi/4 \right], \quad \Omega > 0, \quad t'' < x < \bar{t}',$$

$$(3.23.1b)$$

where $\delta'(\bar{k})$ is given by (3.22.10a) with k replaced by $\bar{k}$. Since $|q(\bar{x}_1)| = |q(x_1)|$ we have according to (3.22.10b) and (3.22.10a)

$$\delta''(k) = \tfrac{1}{2}\tilde{\phi} + \tfrac{1}{2}\arctan\exp(-K)$$

$$- \arctan \frac{k/|q(x_1)|}{[\exp(K) + \gamma]k/|q(x_1)| + [\exp(K) - \gamma]} \quad (\text{mod } \pi), \quad (3.23.2a)$$

and

$$\delta'(\bar{k}) = \tfrac{1}{2}\tilde{\phi} + \tfrac{1}{2}\arctan\exp(-K)$$

$$- \arctan \frac{\bar{k}/|q(x_1)|}{[\exp(K) + \gamma]\bar{k}/|q(x_1)| - [\exp(K) - \gamma]} \quad (\text{mod } \pi), \quad (3.23.2b)$$

respectively. From the treatment in Section 3.22 one knows that when k and $\bar{k}$ are equal to zero or infinity, (2.23.2a,b) are also valid when the barriers are thin. For other values of k and $\bar{k}$ it is required that the barriers be thick. Identifying the expressions in (3.23.1a) and (3.23.1b), we obtain the quantization condition

$$\cos\left[\left|\int_{(t'')}^{x} q(z)dz\right| + \delta''(k) - \pi/4\right] = \cos\left[\left|\int_{(\bar{t}')}^{x} q(z)dz\right| + \delta'(\bar{k}) - \pi/4\right], \quad (3.23.3)$$

which we rewrite as

$$\cos\left[\left|\int_{(t'')}^{x} q(z)dz\right| + \delta''(k) - \pi/4\right] = \cos\left[\left|\int_{(t'')}^{x} q(z)dz\right| - L - \delta'(\bar{k}) + \pi/4\right],$$

$$(3.23.4)$$

where

$$L = \left|\int_{(t'')}^{(\bar{t}')} q(z)dz\right|. \quad (3.23.5)$$

From (3.23.4) we obtain

$$L = \pi/2 - [\delta''(k) + \delta'(\bar{k})] \quad (\text{mod } 2\pi). \quad (3.23.6)$$

Considering the case $\bar{k} = -k$, we obtain from (3.23.2a,b)

$$\delta''(k) + \delta'(\bar{k}) = \delta''(k) + \delta'(-k) = \tilde{\phi} + \arctan\exp(-K)$$

$$- 2\arctan \frac{k/|q(x_1)|}{[\exp(K) + \gamma]k/|q(x_1)| + [\exp(K) - \gamma]} \quad (\text{mod } \pi).$$

$$(3.23.7)$$

When the barriers are thick, we obtain from (3.23.7) approximately (cf. (2.4.8c) with $|w(x_1)| = K/2$)

$$\delta''(k) + \delta'(\bar{k}) = \tilde{\phi} + \exp(-K) - 2\frac{k/|q(x_1)|}{[\exp(K) + \gamma]k/|q(x_1)| + [\exp(K) - \gamma]}$$

$$\approx \tilde{\phi} - \frac{[k/|q(x_1)| - 1]\exp(-K)}{[1 + \gamma\exp(-K)]k/|q(x_1)| + [1 - \gamma\exp(-K)]} \quad (\text{mod } \pi).$$

$$(3.23.8)$$

We insert (3.23.8) into (3.23.6) and obtain the quantization condition

$$L = (s + 1/2)\pi - \tilde{\phi} + \frac{[k/|q(x_1)| - 1]\exp(-K)}{[1 + \gamma\exp(-K)]k/|q(x_1)| + [1 - \gamma\exp(-K)]},$$

$$(3.23.9)$$

where s is an integer. From (2.4.8c) with $|w(x_1)| = K/2$ it follows that $\gamma\exp(-K)$ is small compared with unity. When k is not too close to $-|q(x_1)|$, one can therefore neglect $\gamma\exp(-K)$ in (3.23.9). Since the right-hand side of (3.23.9) increases monotonically as k increases, L and hence the energy increase as k increases. Increase of k (> 0) corresponds to the compression of an atom, and it is known that the energy increases when the atom is compressed; see Sections 3.29 and 3.30. Our conclusion is thus consistent with this fact. When k increases from 0 to $+\infty$, it is seen from (3.23.9) that L increases from approximately $(s + 1/2)\pi - \tilde{\phi} - \exp(-K)$ to approximately $(s + 1/2)\pi - \tilde{\phi} + \exp(-K)$.

3.24 Dependence of the phase of the wave function in a classically allowed region on the position of the point x_1 in an adjacent classically forbidden region where the boundary condition $\psi(x_1) = 0$ is imposed

For a real, smooth potential, consider a simple, well-isolated turning point t and impose the boundary condition that the wave function be equal to zero at the point x_1 in the classically forbidden region. Investigate, with the aid of formulas derived in Section 3.21, the change of the phase of the wave function at a fixed point x_2 in the classically allowed region, when the position of the point x_1 changes. Like the solution in Section 3.21, the solution of the present problem is one of the most important applications of the F-matrix method and the parameterization of the F-matrix, since the problem cannot be solved within the framework of asymptotic analysis.

Solution. Simultaneously considering the cases in Fig. 2.4.1 and Fig. 2.4.2, with $w(z)$ defined by (2.4.1), we have

$$q^{1/2}(x_1) = \exp(\pm i\pi/4)|q^{1/2}(x_1)|, \tag{3.24.1a}$$

$$w(x_1) = -i|w(x_1)|, \tag{3.24.1b}$$

$$q^{1/2}(x_2) = |q^{1/2}(x_2)|, \tag{3.24.2a}$$

$$w(x_2) = \pm|w(x_2)|, \tag{3.24.2b}$$

where the upper and the lower signs are appropriate for the cases in Fig. 2.4.1 and Fig. 2.4.2, respectively. According to (3.21.1) one obtains formulas pertinent to the boundary condition $\psi(x_1) = 0$ by putting $k = \infty$ and hence, according to (3.21.4), $\kappa(k) = -1$. From (3.21.5) we therefore obtain

$$a_1(x_2) = -\{\exp[-2i\,w(x_1)]F_{22}(x_1, x_2) + F_{12}(x_1, x_2)\}a_2(x_1). \tag{3.24.3}$$

As in Section 3.21 we assume the wave function to be real on the real z-axis. According to (3.21.8), with $\delta(k)$ replaced by $\delta(x_1, x_2)$, and (3.21.9a,b) the wave function at the point x_2 in the classically allowed region is

$$\psi(x_2) = \Omega(x_2)|q^{-1/2}(x_2)| \cos[|w(x_2)| + \delta(x_1, x_2) - \pi/4], \tag{3.24.4}$$

where

$$\Omega(x_2) = 2|a_1(x_2)|, \tag{3.24.5a}$$

$$\delta(x_1, x_2) = \pi/4 \pm \arg a_1(x_2). \tag{3.24.5b}$$

We shall now assume that the point x_1 moves, while the point x_2 and the coefficient $a_2(x_1)$ are kept fixed (see the explanation above (3.21.11)), and we shall calculate useful expressions for the derivatives with respect to x_1 of $|a_1(x_2)|$ and $\arg a_1(x_2)$. To do this we first note that

$$\ln a_1(x_2) = \ln|a_1(x_2)| + i \arg a_1(x_2) \tag{3.24.6}$$

and hence

$$\frac{\partial a_1(x_2)/\partial x_1}{a_1(x_2)} = \frac{\partial|a_1 x_2)|/\partial x_1}{|a_1(x_2)|} + i\frac{\partial[\arg a_1(x_2)]}{\partial x_1}, \tag{3.24.7}$$

i.e.,

$$\frac{\partial|a_1(x_2)|/\partial x_1}{|a_1(x_2)|} = \text{Re}\frac{\partial a_1(x_2)/\partial x_1}{a_1(x_2)}, \tag{3.24.8a}$$

$$\frac{\partial[\arg a_1(x_2)]}{\partial x_1} = \text{Im}\frac{\partial a_1(x_2)/\partial x_1}{a_1(x_2)}. \tag{3.24.8b}$$

When $a_2(x_1)$ is kept fixed we obtain from (3.24.3) and (2.3.14b)

$$\frac{\partial a_1(x_2)/\partial x_1}{a_1(x_2)} = -\frac{2iq(x_1)\exp[-2iw(x_1)]\,F_{22}(x_1,x_2)}{\exp[-2iw(x_1)]\,F_{22}(x_1,x_2)+F_{12}(x_1,x_2)}. \qquad (3.24.9)$$

Introducing into (3.24.9) the expressions for the F-matrix elements in terms of the parameters α, β and γ according to (2.4.9b,d), and noting that according to (3.24.1a,b)

$$iq(x_1) = \mp|q(x_1)|, \qquad (3.24.10a)$$
$$iw(x_1) = |w(x_1)|, \qquad (3.24.10b)$$

we obtain

$$\frac{\partial a_1(x_2)/\partial x_1}{a_1(x_2)} = \frac{\pm 2|q(x_1)|\left(\gamma \mp \dfrac{i}{2\alpha^2}\right)\exp[-2|w(x_1)|]}{1+\gamma\exp[-2|w(x_1)|]\mp\dfrac{i}{2\alpha^2}\exp[-2|w(x_1)|]}$$

$$= \pm 2|q(x_1)|\frac{\gamma\exp[-2|w\,(x_1)|]\{1+\gamma\exp[-2|w(x_1)|]\}+\dfrac{1}{4\alpha^4}\exp[-4|w(x_1)|]}{\{1+\gamma\exp[-2|w(x_1)|]\}^2+\dfrac{1}{4\alpha^4}\exp[-4|w(x_1)|]}$$

$$-i|q(x_1)|\frac{\dfrac{1}{\alpha^2}\exp[-2|w(x_1)|]}{\{1+\gamma\exp[-2|w(x_1)|]\}^2+\dfrac{1}{4\alpha^4}\exp[-4|w(x_1)|]}. \qquad (3.24.11)$$

Recalling the estimates (2.4.8a,c) for α and γ, we see from (3.24.11) that, for the logarithmic derivative of $a_1(x_2)$ with respect to x_1, the real part is only determined to be of the order $|q(x_1)|\mu$, while the imaginary part is given by the approximate expression $-|q(x_1)|\exp[-2|w(x_1)|]$. According to (3.24.8a) and (3.24.11) we are therefore not able to obtain an approximate expression for the logarithmic derivative of $|a_1(x_2)|$ with respect to x_1, but only the estimate

$$\frac{\partial|a_1(x_2)|/\partial x_1}{|a_1(x_2)|} = |q(x_1)|O(\mu), \qquad (3.24.12a)$$

while for $\partial[\arg a_1(x_2)]/\partial x_1$ we obtain by using (3.24.8b), (3.24.11), (2.4.8a,c) and (3.24.10a,b)

$$\frac{\partial[\arg a_1(x_2)]}{\partial x_1} \approx -|q(x_1)|\exp[-2|w(x_1)|] = \pm iq(x_1)\exp[-2iw(x_1)]$$

$$= \mp\frac{1}{2}\frac{\partial}{\partial x_1}\exp[-2iw(x_1)] = \mp\frac{1}{2}\frac{\partial}{\partial x_1}\exp[-2|w(x_1)|], \qquad (3.24.12b)$$

and hence

$$[\arg a_1(x_2)]_{x_1} - [\arg a_1(x_2)]_{\bar{x}_1} \approx \mp\tfrac{1}{2}\{\exp[-2|w(x_1)|] - \exp[-2|w(\bar{x}_1)|]\}.$$
(3.24.13)

From (3.24.5b) and (3.24.13) we obtain

$$\delta(x_1, x_2) - \delta(\bar{x}_1, x_2) = \pm\{[\arg a_1(x_2)]_{x_1} - [\arg a_1(x_2)]_{\bar{x}_1}\}$$
$$\approx \tfrac{1}{2}\exp[-2|w(\bar{x}_1)|] - \tfrac{1}{2}\exp[-2|w(x_1)|]. \quad (3.24.14)$$

From (3.24.14) it is seen how the phase of the wave function at the fixed point x_2 changes, when the point, where the boundary condition that the wave function be equal to zero is imposed, moves from $\bar{x}_1$ to x_1. Formula (3.24.14) forms the basis for the simple treatment in Section 3.29 of the displacement of the energy levels due to the compression of an atom.

Using (2.4.9b,d) and the estimates (2.4.8a–c) of α, β and γ along with (2.4.6), we obtain from (3.24.3)

$$a_1(x_2) = \mp i[1 + O(\mu)]a_2(x_1), \quad (3.24.15)$$

where $O(\mu)$ denotes a complex quantity at the most of the order of magnitude μ. Insertion of (3.24.15) into (3.24.5b) yields

$$\delta(x_1, x_2) = -\pi/4 \pm \arg a_2(x_1) + O(\mu). \quad (3.24.16)$$

Although, according to (3.24.16), we cannot determine the magnitude of the phase more accurately than to within a μ-quantity, we can determine its *exponentially small change* according to (3.24.14), when the point x_1, where the boundary condition $\psi(x_1) = 0$ is imposed, is moved.

As a particular case of (3.24.14), if the classically forbidden region extends to X_1 and $|w(\bar{x}_1)| \to \infty$ as $\bar{x}_1 \to X_1$, we obtain the formula

$$\delta(x_1, x_2) - \delta(X_1, x_2) = -\tfrac{1}{2}\exp[-2|w(x_1)|], \quad (3.24.17)$$

which can also be derived in another way, as we shall now show. Consider the case in Fig. 2.4.2. If the barrier extends to X_1 and $|w(X_1)| = \infty$, and if the boundary condition is that $\psi(X_1) = 0$, the wave function is proportional to $f_2(x)$ when x lies sufficiently far to the right of t. Hence the boundary condition at X_1 can be replaced by the boundary condition that the logarithmic derivative of the wave function at the point x_1 in Fig. 2.4.2 be (cf. (2.3.1b,c) and (3.24.10a) with the lower sign)

$$k = \frac{\psi'(x_1)}{\psi(x_1)} = \frac{f_2'(x_1)}{f_2(x_1)} = -iq(x_1) - \frac{q'(x_1)}{2q(x_1)} = -|q(x_1)|\left\{1 - \frac{1}{2}d[1/|q(x_1)|]/dx_1\right\},$$
(3.24.18)

which according to (3.21.4) with the lower sign means that

$$\kappa = 0. \tag{3.24.19}$$

If the boundary condition is instead $\psi(x_1) = 0$, which means that $k = \infty$, we obtain from (3.21.4)

$$\kappa = -1. \tag{3.24.20}$$

According to (3.24.19) and (3.24.20) the displacement from X_1 to x_1 of the point, where it is required that the wave function be equal to zero, thus corresponds to the replacement of $\kappa = 0$ by $\kappa = -1$. From (3.21.10) and the estimates (2.4.8a,c) of α and γ we therefore obtain the approximate formula

$$\delta[\psi(x_1) = 0] - \delta[\psi(X_1) = 0] = \arctan\left\{-\tfrac{1}{2}\exp[-2|w(x_1)|]\right\}. \tag{3.24.21}$$

Since in the derivation of this formula the denominator in the last term of (3.21.10) has been approximated by 2, it is not significant to give the right-hand side of (3.24.21) more accurately than as the first term of the Taylor expansion of arctan. One thus obtains (3.24.17).

3.25 Phase-shift formula

Derive a formula for the phase shift δ_l of a radial partial wave with the angular momentum quantum number l for the case of a monotonically decreasing physical potential $V(r)$, which tends to zero faster than const/r^2 as $r \to +\infty$.

Solution. The radial wave function ψ for a quantal particle in the spherically symmetric physical potential $V(r)$ satisfies the radial Schrödinger equation

$$\frac{d^2\psi}{dr^2} + R(r)\psi = 0, \tag{3.25.1}$$

where

$$R(r) = k^2 - (2m/\hbar^2)V(r) - l(l+1)/r^2, \tag{3.25.2a}$$
$$k = (2mE)^{1/2}/\hbar. \tag{3.25.2b}$$

We choose the square of the base function to be

$$Q^2(r) = k^2 - (2m/\hbar^2)V(r) - (l+1/2)^2/r^2. \tag{3.25.3}$$

With this choice of base function the arbitrary-order phase-integral approximation remains valid close to the origin; see Sections 3.1 and 3.2. Since $V(r)$ is monotonically decreasing and tends to zero as r tends to infinity, there is a single turning point $t(> 0)$ when $E > 0$. The radial wave function that is equal to zero at the origin is (except for a constant factor) close to the origin given by the phase-integral formula

$$\psi = |q^{-1/2}(r)| \exp \left[- \left| \int_{(t)}^{r} q(r) dr \right| \right], \quad r \to 0. \tag{3.25.4}$$

Since $l \neq -1/2$, one finds using the connection formula (3.10.10') that in the region $r > t$ this wave function is

$$\psi = 2q^{-1/2}(r) \cos \left[\int_{(t)}^{r} q(r) dr - \pi/4 \right], \quad r > t, \tag{3.25.5}$$

if $Q(r)$, and hence $q(r)$, is chosen to be positive for $r > t$. For a free particle one has an analogous formula with $V(r) \equiv 0$, i.e.,

$$\psi = q_0^{-1/2}(r) \cos \left[\int_{(t_0)}^{r} q_0(r) dr - \pi/4 \right], \quad r > t_0, \tag{3.25.6}$$

where $q_0(r)$ is the phase integrand when $V(r) \equiv 0$, and t_0 is the corresponding turning point. By definition the phase shift δ_l is the difference between the argument of the cosine in (3.25.5) and the argument of the cosine in (3.25.6) in the limit when $r \to +\infty$ and is thus given by the formula

$$\delta_l = \lim_{r \to +\infty} \left(\int_{(t)}^{r} q(r) dr - \int_{(t_0)}^{r} q_0(r) dr \right). \tag{3.25.7}$$

Denoting the base function when $V(r) \equiv 0$ by $Q_0(r)$, and noting that $t_0 = (l + 1/2)/k$, one obtains in the first-order approximation

$$\int_{t_0}^{r} Q_0(r) dr = \int_{(l+1/2)/k}^{r} [k^2 - (l + 1/2)^2/r^2]^{1/2} dr$$

$$= [(kr)^2 - (l + 1/2)^2]^{1/2} + (l + 1/2)\{\arcsin[(l + 1/2)/(kr)] - \pi/2\}$$

$$\underset{r \to +\infty}{\sim} kr - (l + 1/2)\pi/2, \quad V(r) \equiv 0. \tag{3.25.8}$$

Higher-order terms in $q_0(r)$ do not contribute to the phase integral $\int_{(t_0)}^{r} q_0(r) dr$ in the limit $r \to +\infty$, and hence

$$\int_{(t_0)}^{r} q_0(r)dr \sim kr - (l + 1/2)\pi/2, \quad r \to +\infty, \tag{3.25.9}$$

for an arbitrary order of approximation. The phase-shift formula (3.25.7) can therefore be written as

$$\delta_l = \lim_{r \to +\infty} \left[\int_{(t)}^{r} q(r)dr - kr + (l + 1/2)\pi/2 \right]. \tag{3.25.10}$$

With the choice (3.25.3) of $Q^2(r)$ this formula gives the correct value $\delta_l = 0$ of the phase shift when $V(r) \equiv 0$.

Using (2.2.11), (2.2.7a) and (3.25.3), we can alternatively write (3.25.7) as

$$\delta_l = \lim_{r \to +\infty} \left\{ \int_{t}^{r} Q(r)dr - \int_{t_0}^{r} [k^2 - (l + 1/2)^2/r^2]^{1/2} dr \right\}$$

$$+ \sum_{n=1}^{N} \left\{ \int_{(t)}^{+\infty} Y_{2n} Q(r)dr - \int_{(t_0)}^{+\infty} [Y_{2n} Q(r)]_{V(r)\equiv 0} dr \right\}. \tag{3.25.11}$$

It may be preferable to use (3.25.11) instead of (3.25.10), since errors in $q(r)$ and $q_0(r)$ may cancel in (3.25.11).

If the potential $V(r)$ has a stronger singularity than $const/r^2$ at the origin, the phase-integral approximation is according to Section 3.2 also valid close to the origin with the choice $Q^2(r) = R(r)$. However, it may also be convenient to choose $Q^2(r)$ according to (3.25.3) in that case in order to obtain $\delta_l = 0$ from (3.25.10) when $V(r) \equiv 0$.

3.26 Distance between near-lying energy levels in different types of physical systems, expressed either in terms of the frequency of classical oscillations in a potential well or in terms of the derivative of the energy with respect to a quantum number

Assume that the near-lying energy values E_1 and E_2 are obtained from the quantization conditions

$$L(E_1) = (s + \Delta + \Delta_1)\pi, \tag{3.26.1}$$

$$L(E_2) = (s + \Delta + \Delta_2)\pi, \tag{3.26.2}$$

where

$$L(E) = \int\limits_{(t')}^{(t'')} q(z)dz \tag{3.26.3}$$

is the phase integral associated with a potential well between the turning points t' and t'' $(> t')$, and Δ_1 and Δ_2 are unspecified, small constants. We shall in later sections derive quantization conditions that are particular cases of (3.26.1) and (3.26.2). For the discussion of those results it is convenient to treat the problem in this section now in order to avoid repetition of essentially the same calculations several times in following sections. $E_1 - E_2$ may be the difference between two neighbouring energy eigenvalues pertaining to a single-well potential, the displacement of an energy level due to the compression of an atom, the energy splitting of two fine structure levels of a double- or multi-well potential, the distance between two adjacent energies for which a symmetric double-hump potential is completely transparent, the width of an energy band in a periodic potential, or the imaginary part of the energy of a quasi-stationary state.

Show that $E_1 - E_2$ can be expressed either in terms of the frequency of classical oscillations in the potential well associated with $-Q^2(r)$ and the quantization condition

$$L(E_0) = (s + \Delta)\pi, \tag{3.26.4}$$

or in terms of dE_0/ds.

Solution. When $E - E_0$ is sufficiently small, we have the approximate formula

$$L(E) = L(E_0) + [dL(E)/dE]_{E=E_0}(E - E_0). \tag{3.26.5}$$

Taking the difference of (3.26.1) and (3.26.2), and using (3.26.5), we obtain approximately

$$E_1 - E_2 = \frac{(\Delta_1 - \Delta_2)\,\pi}{[dL(E)/dE]_{E=E_0}}. \tag{3.26.6}$$

Using the first-order expression for $L(E)$, we obtain if $Q^2(z) = 2m[E - V(z)]/\hbar^2$

$$\left[\frac{dL(E)}{dE}\right]_{E=E_0} = \left[\frac{d}{dE}\int_{t'}^{t''} Q(z)dz\right]_{E=E_0} = \left(\frac{d}{dE}\int_{t'}^{t''} \{2m[E-V(z)]/\hbar^2\}^{1/2}dz\right)_{E=E_0}$$

$$= \frac{1}{\hbar}\int_{t'_0}^{t''_0} \frac{dz}{\{2[E_0 - V(z)]/m\}^{1/2}} = \frac{\tau}{2\hbar} = \frac{\pi\tau}{h} = \frac{\pi}{h\nu}, \tag{3.26.7}$$

where

$$\tau = 2 \int\limits_{t_0'}^{t_0''} \frac{dz}{\{2[E_0 - V(z)]/m\}^{1/2}} \qquad (3.26.8)$$

is the time for a full oscillation between the turning points of a classical particle with energy E_0, and $\nu = 1/\tau$ is the corresponding frequency. Inserting (3.26.7) into (3.26.6), we obtain

$$E_1 - E_2 = h(\Delta_1 - \Delta_2)/\tau = h\nu(\Delta_1 - \Delta_2). \qquad (3.26.9)$$

We shall now obtain an alternative formula for $E_1 - E_2$. From (3.26.4) it follows that approximately

$$\frac{dL(E_0)}{dE_0} = \pi \frac{ds}{dE_0} = \frac{\pi}{dE_0/ds}, \qquad (3.26.10)$$

and inserting (3.26.10) into (3.26.6), we obtain

$$E_1 - E_2 \approx \frac{dE_0}{ds}(\Delta_1 - \Delta_2). \qquad (3.26.11)$$

3.27 Arbitrary-order quantization condition for a particle in a single-well potential, derived on the assumption that the classically allowed region is broad enough to allow the use of a connection formula

Derive the arbitrary-order quantization condition for a non-relativistic quantal particle in a broad single-well potential with the turning points t' and t'' $(> t')$.

Solution. We assume the boundary conditions to require that the wave function be equal to zero for $z = a$ and $z = b$. One easily obtains the quantization condition by tracing bound-state wave functions from both a and b into the potential well and matching the resulting two wave functions there. We choose the base function $Q(x)$ and also $Q^{1/2}(x)$ to be positive in the interval $t' < x < t''$, which implies that $q^{1/2}(x)$ is also positive in the same interval, except possibly for small regions close to the turning points t' and t'' $(> t')$, where $q(x)$ may have first-order zeros; see the text below (2.2.11). The boundary conditions $\psi(a) = 0$ and $\psi(b) = 0$ are fulfilled by

$$\psi(x) = D' |q^{-1/2}(x)| \exp\left[-\left| \int\limits_{(t')}^{x} q(z)dz \right| \right] \qquad \text{for} \quad x < t', \qquad (3.27.1a)$$

$$\psi(x) = D'' |q^{-1/2}(x)| \exp\left[-\left| \int\limits_{(t'')}^{x} q(z)dz \right| \right] \qquad \text{for} \quad x > t'', \qquad (3.27.1b)$$

if the exponentials tend to zero as x tends to a and to b, respectively. By means of the connection formula (3.10.10') we obtain

$$\psi(x) = 2D'q^{-1/2}(x)\cos\left[\int_{(t')}^{x} q(z)dz - \pi/4\right] \quad \text{for} \quad t' < x < t'', \quad (3.27.2a)$$

$$\psi(x) = 2D''q^{-1/2}(x)\cos\left[\int_{x}^{(t'')} q(z)dz - \pi/4\right] \quad \text{for} \quad t' < x < t''. \quad (3.27.2b)$$

Fitting the expressions (3.27.2a) and (3.27.2b) for the wave function to each other, we get the equation

$$D'\cos\left[\int_{(t')}^{x} q(z)dz - \pi/4\right] = D''\cos\left[\int_{x}^{(t'')} q(z)dz - \pi/4\right], \quad (3.27.3)$$

which we rewrite as

$$D'\cos\left[\int_{(t')}^{x} q(z)dz - \pi/4\right] = D''\cos\left[\int_{(t')}^{x} q(z)dz - L + \pi/4\right], \quad (3.27.4)$$

where

$$L = \int_{(t')}^{(t'')} q(z)dz. \quad (3.27.5)$$

From (3.27.4) we obtain

$$L = (s + 1/2)\pi, \quad (3.27.6a)$$

$$D' = (-1)^s D'', \quad (3.27.6b)$$

where s is an integer, which because of (3.27.5) must be non-negative.

In the $(2N + 1)$th-order approximation we can with the aid of (2.2.11), (2.2.8) and (2.2.4) write (3.27.5) as

$$L = \sum_{n=0}^{N} L^{(2n+1)}, \quad (3.27.7a)$$

$$L^{(2n+1)} = \int_{(t')}^{(t'')} Y_{2n}Q(z)dz = \int_{(t')}^{(t'')} Z_{2n}Q(z)dz, \quad (3.27.7b)$$

where Y_{2n} and Z_{2n} are given by (2.2.7a–d) and (2.2.9a–d). Recalling the meaning of the short-hand notation for the contour integrals, explained in connection with (2.2.14) and (2.2.15), we can also write (3.27.7b) as

$$L^{(2n+1)} = \frac{1}{2}\int_{\Lambda} Y_{2n}Q(z)dz = \frac{1}{2}\int_{\Lambda} Z_{2n}Q(z)dz, \qquad (3.27.7b')$$

where $Q(z)$ is positive on the upper lip of the cut between t' and t'', and Λ is a closed contour, encircling t' and t'', on which the integration is performed in the negative direction.

Choosing $Q^2(z) = R(z)$ and considering two adjacent energy levels E_s and E_{s+1}, we have, with the notation in Section 3.26, $\Delta = 1/2$, $\Delta_1 = 1$ and $\Delta_2 = 0$, and from (3.26.9) and (3.26.11) we obtain the alternative approximate formulas $E_{s+1} - E_s = h\nu$ and $E_{s+1} - E_s = dE_s/ds$.

3.28 Arbitrary-order quantization condition for a particle in a single-well potential, derived without the assumption that the classically allowed region is broad

Derive the arbitrary-order quantization condition for a non-relativistic quantal particle in a single-well potential with the turning points t' and $t''(>t')$ by tracing a bound-state solution in the complex plane along paths on which the absolute value of $\exp[iw(z, E)]$ is monotonic, and using the fact that the wave function is single-valued.

Solution. Using obvious notation, we consider the Schrödinger equation

$$d^2\psi/dz^2 + R(z, E)\psi = 0, \qquad (3.28.1a)$$

$$R(z, E) = (2m/\hbar^2)[E - V(z)], \qquad (3.28.1b)$$

where $V(z)$ may be the actual physical potential or an effective potential. Thus, if we are concerned with the radial Schrödinger equation, $V(z)$ also includes the centrifugal term. We assume that $V(z)$ is a single-well potential that is regular analytic in the region of the complex z-plane under consideration. The points where one imposes the boundary conditions that the wave function be equal to zero are denoted by a and $b\,(>a)$. For a radial problem one has thus $a = 0$ and $b = +\infty$. The eigenfunction (real on the real z-axis) of a bound state with the quantum number s and the energy E_s is denoted by $\psi(z, E_s)$ and is associated with the turning points t' and $t''(>t')$.

The complex z-plane is assumed to be cut along the real z-axis from a to t' as well as between the two relevant turning points t' and t''. Other zeros and singularities of $Q^2(z, E)$, as well as singularities of $R(z, E)$, are assumed to lie far away from the classically allowed region between t' and t''. We choose $Q^2(z, E)$ to be positive on the upper lip of the cut between t' and t''. From the complex z-plane we exclude a certain region around the classically allowed part of the real z-axis, i.e., the real axis between t' and t'', and also a remote region containing the further zeros and the singularities of $Q^2(z, E)$, as well as the singularities of $R(z, E)$, which may possibly exist, and consider a certain band (dashed in Fig. 3.28.1(c)) encircling the classically allowed region. Whether the bottom of the potential is approximately parabolic or not, it follows from the general behaviour in the complex z-plane of the anti-Stokes lines, i.e., the level lines for the absolute value of $\exp[iw(z, E)]$, where $w(z, E) = \int^z q(z, E)dz$ with an unspecified lower limit of integration, that, in this band there is a region L to the left in which any point z can be reached on a path from a (proceeding first along the real z-axis and then in the region L) along which the absolute value of $\exp[-iw(z, E)]$ increases monotonically as one moves from a to z. Similarly there is a region R to the right in which any point z can be reached on a path from b (proceeding first along the real z-axis and then in the region R) along which the absolute value of $\exp[-iw(z, E)]$ increases monotonically as one moves from b to z. For L and R there is a common region above as well as below the real axis; see Fig. 3.28.1. Using results obtained in Section 3.6, one realizes that, even if the energy does not correspond to a bound state, a solution ψ_L of the differential equation (3.28.1a,b) tending to zero as $z \to a$ is, except for a constant factor, represented in the region L by $q^{-1/2}(z, E)\exp[-iw(z, E)]$, if the classically forbidden region in which a lies is thick enough. Similarly, a solution ψ_R tending to zero as $z \to b$ is, except for a constant factor, in the region R also represented by $q^{-1/2}(z, E)\exp[-iw(z, E)]$, if the classically forbidden region in which b lies is thick enough. When the energy is *not* equal to an eigenvalue, $\psi_L \to \infty$ as $z \to b$ and $\psi_R \to \infty$ as $z \to a$, and the solutions ψ_L and ψ_R are thus quite different and linearly independent. Yet they are both (except for a constant factor) represented by the same phase-integral function $q^{-1/2}(z, E)\exp[-iw(z, E)]$ in the common parts of the regions L and R; see Fig. 3.28.1. The eigenfunction $\psi(z, E_s)$ corresponding to the bound state with the energy E_s tends to zero as $z \to a$ and also as $z \to b$ and therefore, except for a normalization factor, can be approximately written as

$$\psi(z, E_s) = q^{-1/2}(z, E_s)\exp[-iw(z, E_s)] \tag{3.28.2}$$

in the whole band consisting of the regions L and R in Fig. 3.28.1(c).

Since the exact wave function is single-valued, the asymptotic solution (3.28.2) must also be single-valued when z moves one turn around both turning points. However, such a circulation changes the right-hand side of (3.28.2) due to the facts that $q(z, E_s)$ has opposite signs on the lower and the upper edges of the cut between

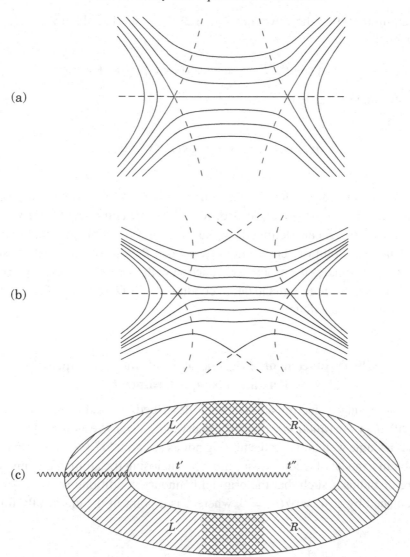

Figure 3.28.1 The full curves in (a), which refers to the case of a harmonic oscillator, and in (b), which refers to the case of a purely quartic oscillator, are anti-Stokes lines, i.e., level lines for the absolute value of $\exp[i\,w(z, E)]$. The dashed lines are Stokes lines, i.e., level lines for constant phase of $\exp[i\,w(z, E)]$. In (c), where there is a cut (wavy line) to the left of t' and between the turning points t' and t'', the regions L (indicated by ////) and R (indicated by \\\\) are shown.

t' and t'' and that $q^{-1/2}(z, E)$ changes sign. We assume that z moves from a point $x + i0$ with $x < t'$ on the upper lip of the cut into the dashed band in Fig. 3.28.1(c) and then proceeds in the band a full turn in the negative direction and finally leaves the band and continues to the point $x - i0$ with $x < t'$ on the lower lip of the cut.

The requirement that the wave function (3.28.2) be single-valued when z moves in this way yields

$$i[w(x - i0, E_s) - w(x + i0, E_s)] = i\pi + 2si\pi, \tag{3.28.3}$$

i.e., according to (2.3.1c)

$$\int_{(t')}^{(t'')} q(z, E_s)dz = (s + 1/2)\pi, \quad s = 0, 1, 2, \ldots. \tag{3.28.4}$$

For the harmonic oscillator the turning points t' and t'' are the only transition points, and there are no singularities of $R(z, E)$ in the complex plane. By choosing $Q^2(z, E) = R(z, E)$ and letting the radius of a circle on which (3.28.2) represents the bound-state solution tend to infinity, we can therefore achieve the result that (3.28.2) approximates the exact wave function with as great an accuracy as desired. Hence the phase-integral quantization condition with $Q^2(z, E) = R(z, E)$ is exact for the harmonic oscillator.

3.29 Displacement of the energy levels due to compression of an atom (simple treatment)

An ordinary single-electron atom or ion is described by a radial Schrödinger equation with the wave function $\psi(r)$ on the interval $(0, +\infty)$. We assume that the potential has a single well and that the turning points are t' and t'' ($> t'$); see Fig. 3.29.1. We obtain a model of a compressed atom by enclosing the atom in an impenetrable sphere of radius a such that the original boundary conditions $\psi(0) = \psi(\infty) = 0$ are replaced by $\psi(0) = \psi(a) = 0$, where a is a point in the classically forbidden

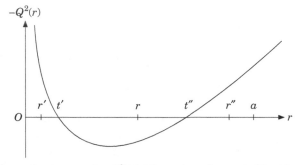

Figure 3.29.1 Schematic picture of $-Q^2(r)$. The points r', r and r'' in this figure are referred to in Section 3.30.

region to the right of t''. Use (3.24.17) to derive a formula for the shifts of the energy levels due to the compression of the atom.

The use of a rigorous procedure is particularly important in the present problem. Since one is looking for a very subtle, in fact 'exponentially small', effect, it is essential to have complete control of the calculations.

Solution. The energy shifts are exponentially small, and although various different methods had been used, including purely numerical ones, in a great number of papers, there existed no general agreement as to the effects considered, and no analytical formula for the energy shifts, until such a formula was derived by Fröman, Yngve and Fröman (1987). The derivation involved rather cumbersome and non-trivial manipulations with the F-matrix in order to show that one can obtain an analytical formula that, if the compression is not too drastic, yields the energy shift with only a small relative error. By means of the parameterized formulas presented in Section 2.4, and with the aid of (3.24.17), we can now derive the formula for the energy shifts in question in a considerably simpler way.

We choose the square of the base function as in (3.1.10), i.e.,

$$Q^2(r) = R(r) - 1/(4r^2). \tag{3.29.1}$$

We assume that the phase integral tends to infinity as r tends to zero. Recalling (2.4.23) with $C(0) = 0$ and $D(0) = 1$, (2.4.26) and (2.4.28a,b) with $\alpha = \alpha'$, $\beta = \beta'$ and $\gamma = \gamma'$, we realize that the requirement that the radial wave function be equal to zero for $r = 0$ gives

$$\psi(r) = 2\alpha' \left| q^{-1/2}(r) \right| \cos \left[\left| \int_{(t')}^{r} q(z)dz \right| - \beta' - \pi/4 \right], \quad t' < r < t'', \tag{3.29.2}$$

where according (2.4.8b) β' is small compared with unity and independent of a. Using the above-mentioned formulas from Section 2.4 but now with $\alpha = \alpha''(a)$, $\beta = \beta''(a)$ and $\gamma = \gamma''(a)$, we find that the requirement that the wave function be equal to zero for $r = a$ gives

$$\psi(r) = \pm\Omega'' \left| q^{-1/2}(r) \right| \cos \left[\left| \int_{(t'')}^{r} q(z)dz \right| + \delta''(a) - \pi/4 \right], \quad t' < r < t'',$$

$$\tag{3.29.3}$$

where Ω'' is a positive amplitude, for which it is not necessary to give the explicit expression, and $\delta''(a)$ is a small a-dependent phase to which we shall return later.

We rewrite (3.29.3) into the form

$$\psi(r) = \pm\Omega''|q^{-1/2}(r)| \cos\left[\left|\int_{(t')}^{r} q(z)dz\right| - L - \delta''(a) + \pi/4\right], \quad t' < r < t'',$$

$$(3.29.4)$$

where

$$L = \left|\int_{(t')}^{(t'')} q(z)dz\right|.$$

$$(3.29.5)$$

Explicit expressions for Ω'' and $\delta''(a)$ can be obtained by using formulas in Section 3.30, but they are not needed in the present problem.

Identifying (3.29.4) with (3.29.2), we obtain

$$L = (s + 1/2)\pi + \beta' - \delta''(a),$$

$$(3.29.6)$$

where s is an integer, which must be non-negative, since L is positive, and β' and $\delta''(a)$ are small. If the energy level corresponding to the quantum number s is denoted by $E_s(a)$, we have according to (3.29.6) the quantization conditions

$$L(E_s(a)) = (s + 1/2)\pi + \beta' - \delta''(a),$$

$$(3.29.7a)$$

$$L(E_s(\infty)) = (s + 1/2)\pi + \beta' - \delta''(\infty),$$

$$(3.29.7b)$$

and hence, since β' does not depend on a,

$$L(E_s(a)) - L(E_s(\infty)) = \delta''(\infty) - \delta''(a).$$

$$(3.29.8)$$

If $\int_{(t'')}^{\infty} q(z)dz = \infty$, we have according to (3.24.17) approximately

$$\delta''(\infty) - \delta''(a) = \tfrac{1}{2}\exp(-2K),$$

$$(3.29.9)$$

where

$$K = \left|\int_{(t'')}^{a} q(z)dz\right|.$$

$$(3.29.10)$$

From (3.29.8) and (3.29.9) we get

$$L(E_s(a)) - L(E_s(\infty)) = \tfrac{1}{2}\exp(-2K).$$

$$(3.29.11)$$

This formula, together with (3.29.5), shows that the compression causes a slight shift upwards of each energy level.

From (3.29.11) we obtain approximately

$$\frac{\partial L}{\partial E}[E_s(a) - E_s(\infty)] \approx \tfrac{1}{2}\exp(-2K), \tag{3.29.12}$$

and from the quantization condition (3.29.6) we obtain $(\partial L/\partial E)dE/ds \approx \pi$, i.e.,

$$\frac{\partial L}{\partial E} \approx \frac{\pi}{dE/ds} \approx \frac{\pi}{E_{s+1} - E_s}. \tag{3.29.13}$$

Combining this formula with (3.29.12), we obtain the formula

$$E_s(a) \approx E_s(\infty) + \frac{\exp(-2K)}{2\pi}(E_{s+1} - E_s), \tag{3.29.14}$$

which yields the change in energy due to the compression. We refer the reader to Fröman, Yngve and Fröman (1987) for references to earlier papers and for a numerical illustration of the accuracy of the displacements of the energy levels that have been calculated in this problem.

3.30 Displacement of the energy levels due to compression of an atom (alternative treatment)

Solve the problem in Section 3.29 by application of the parameterized formulas presented in Section 2.4, but without the use of (3.24.17), in order to explain the strange fact that one obtains the correct displacements of the energy levels by using one connection formula in the allowed direction, but another connection formula *erroneously* in the forbidden direction.

Solution. We choose again the square of the base function as in (3.29.1). As in Section 3.29 we assume that the phase-integral tends to infinity as r tends to zero. At the point r' ($< t'$) close to the origin (see Fig. 3.29.1) the wave function is (apart from a constant factor)

$$\psi(r') = |q^{-1/2}(r')|\exp\left[-\left|\int_{(t')}^{r'} q(z)dz\right|\right], \quad r' < t', \tag{3.30.1}$$

and at the point r in the classically allowed region between t' and t'' this wave function is, according to (2.4.23) with $C(0) = 0$ and $D(0) = 1$, (2.4.26) and (2.4.28a,b) with $\alpha = \alpha'$, $\beta = \beta'$ and $\gamma = \gamma'$ (cf. (3.29.2)),

$$\psi(r) = \alpha'\exp[-i(\pi/4+\beta')]|q^{-1/2}(r)|\exp\left[+i\left|\int_{(t')}^{r}q(z)dz\right|\right] + \text{compl conj}$$

$$= \alpha'\exp[-i(\pi/4+\beta')]|q^{-1/2}(r)|\exp\left\{+i\left[L-\left|\int_{(t'')}^{r}q(z)dz\right|\right]\right\} + \text{compl conj}$$

$$= \alpha'\exp[-i(L-\pi/4-\beta')]|q^{-1/2}(r)|\exp\left[+i\left|\int_{(t'')}^{r}q(z)dz\right|\right] + \text{compl conj},$$

$$(3.30.2)$$

where L is given by (3.29.5). To prepare for the discussion at the end of the present problem, we rewrite the first and the last expression for $\psi(r)$ in (3.30.2) as follows:

$$\psi(r) = 2\alpha'|q^{-1/2}(r)|\cos\left[\left|\int_{(t')}^{r}q(z)dz\right| - \beta' - \pi/4\right], \qquad (3.30.2')$$

$$\psi(r) = 2\alpha'|q^{-1/2}(r)|\cos\left[\left|\int_{(t'')}^{r}q(z)dz\right| - L + \beta' + \pi/4\right]$$

$$= -2\alpha'\sin(L-\pi/2-\beta')|q^{-1/2}(r)|\cos\left[\left|\int_{(t'')}^{r}q(z)dz\right| + \pi/4\right]$$

$$+ 2\alpha'\cos(L-\pi/2-\beta')|q^{-1/2}(r)|\cos\left[\left|\int_{(t'')}^{r}q(z)dz\right| - \pi/4\right]. \quad (3.30.2'')$$

At the point r'' $(>t'')$ in Fig. 3.29.1 the wave function that at the point r is given by (3.30.2) is, according to (2.4.26), (2.4.23) and (2.4.29a,b) with $\alpha = \alpha''$, $\beta = \beta''$ and $\gamma = \gamma''$,

$$\psi(r'') = \{\alpha''\exp[-i(\pi/4-\beta'')]\alpha'\exp[-i(L-\pi/4-\beta')] + \text{compl conj}\}$$

$$\times |q^{-1/2}(r'')|\exp\left[+\left|\int_{(t'')}^{r''}q(z)dz\right|\right]$$

$$+ \{(\alpha''\gamma'' + i/(2\alpha''))\exp[-i(\pi/4-\beta'')]$$

$$\times \alpha'\exp[-i(L-\pi/4-\beta')] + \text{compl conj}\}$$

$$\times |q^{-1/2}(r'')|\exp\left[-\left|\int_{(t'')}^{r''}q(z)dz\right|\right]$$

$$= -2\alpha'\alpha'' \sin(L - \pi/2 - \beta' - \beta'')\left|q^{-1/2}(r'')\right|\exp\left[+\left|\int_{(t'')}^{r''} q(z)dz\right|\right]$$

$$+\left[\frac{\alpha'}{\alpha''}\cos(L - \pi/2 - \beta' - \beta'') - 2\alpha'\alpha''\gamma'' \sin(L - \pi/2 - \beta' - \beta'')\right]$$

$$\times\left|q^{-1/2}(r'')\right|\exp\left[-\left|\int_{(t'')}^{r''} q(z)dz\right|\right]. \tag{3.30.3}$$

Requiring that $\psi(a) = 0$, we obtain from (3.30.3) with $r'' = a$ the condition

$$\tan(L - \pi/2 - \beta' - \beta'') = \frac{\exp(-2K)}{2(\alpha'')^2[1 + \gamma''\exp(-2K)]}, \tag{3.30.4}$$

where K is defined by (3.29.10). From (3.30.4) we obtain with the aid of the estimates (2.4.8a,c) approximately

$$L = (s + 1/2)\pi + \beta' + \beta'' + \arctan[\exp(-2K)/2], \tag{3.30.5}$$

or, since $\exp(-2K)$ is small compared to unity,

$$L = (s + 1/2)\pi + \beta' + \beta'' + \tfrac{1}{2}\exp(-2K). \tag{3.30.5'}$$

Here we cannot neglect β' and β'', since these quantities in general are of the order of magnitude μ and hence large compared with $\exp(-2K)$, although they are small compared with unity; see (2.4.8b) and (2.4.6). However, we know that β' is indepenent of a, and it is reasonable to assume that the relative change of the small parameter β'', when the point where the boundary condition is imposed moves from $+\infty$ to a, is very small. Writing as in (3.29.7a) $L(E_s(a))$ instead of L, we thus obtain from (3.30.5') the approximate formula

$$L(E_s(a)) - L(E_s(\infty)) = \tfrac{1}{2}\exp(-2K), \tag{3.30.6}$$

which agrees with (3.29.11).

One often encounters, even in well-known textbooks, treatments in which the connection formulas (3.10.10') and (3.12.9'), i.e.,

$$\left|q^{-1/2}(x)\right|\exp\left[-\left|\int_{(t)}^{x} q(z)dz\right|\right] \to 2\left|q^{-1/2}(x)\right|\cos\left[\left|\int_{(t)}^{x} q(z)dz\right| - \pi/4\right],$$

$$\tag{3.30.7a}$$

$$\left|q^{-1/2}(x)\right|\cos\left[\left|\int_{(t)}^{x} q(z)dz\right| + \pi/4\right] \to \left|q^{-1/2}(x)\right|\exp\left[+\left|\int_{(t)}^{x} q(z)dz\right|\right],$$

$$\tag{3.30.7b}$$

are used with their one-directional nature disregarded. This is definitely not allowed, and it is therefore puzzling that in that erroneous way one can sometimes obtain an essentially correct *final* result. To account for this strange fact we shall now explain, with the aid of the parameterized formulas in Section 2.4, why one can obtain the energy shift formula (3.30.6) correctly even though one uses the connection formulas (3.30.7a,b) without regard to their one-directional nature, which makes the formulas in the *intermediate* steps incorrect.

We start from (3.30.1) and obtain with the aid of (3.30.7a) (correctly used) for the wave function in the classically allowed region the formulas (3.30.2′) and (3.30.2″) with $\alpha' = 1$ and $\beta' = 0$. With the aid of (3.30.7a) with its one-directional nature disregarded, and (3.30.7b) (correctly used) we then obtain for the wave function in the classically forbidden region to the right of t'' the formula (3.30.3), but with $\alpha' = \alpha'' = 1$ and $\beta' = \beta'' = \gamma'' = 0$. Requiring that $\psi(a) = 0$, we obtain (3.30.4), but with $\alpha'' = 1$ and $\beta' = \beta'' = \gamma'' = 0$, and hence (3.30.5) and (3.30.5′), but with $\beta' = \beta'' = 0$. With these values of the parameters, the formula (3.30.5′) does not give the correct value for the difference $L - (s + 1/2)\pi$, since $\exp(-2K)$ is in general much smaller than the absolute values of the correct values of β' and β''. From (3.30.5′) with $\beta' = \beta'' = 0$ one then obtains the correct expression (3.30.6) for the displacement of the energy level E_s. In the correct treatment one arrives at (3.30.4) from which one obtains (3.30.6) by using the estimates (2.4.8a,c). In the incorrect treatment one arrives at (3.30.4) with the approximately correct values $\alpha'' = 1$ and $\beta' = \beta'' = 0$ and the erroneous value $\gamma'' = 0$. Since the estimates (2.4.8a–c) are fulfilled for these values, the erroneous treatment gives the correct result (3.30.6).

We shall now discuss in a more general way, when and why one can sometimes obtain an essentially correct result by using the connection formulas (3.30.7a,b) with their one-directional nature disregarded.

The exact expression for a physical quantity (e.g., a change of phase or of energy level), derived with the aid of the exact parameterized formulas in Section 2.4 may be so insensitive to the values of α, β and γ associated with each turning point that these parameters disappear approximately from the expression in question when the estimates (2.4.8a–c) of α, β and γ are used. If this is the case, one obtains an essentially correct approximate *final* formula (which may, however, contain insignificant terms) by putting from the beginning $\alpha = 1$, $\beta = 0$ and *erroneously* $\gamma = 0$, which is equivalent to using the connection formulas (3.30.7a,b) with their one-directional nature disregarded. However, *intermediate* steps in the calculations may be quite erroneous. One can thus obtain an essentially correct *final* result, although *intermediate* steps in the calculations may be quite wrong.

On the other hand, *if* one does *not* obtain for the physical quantity in question, with the aid of the parameterized formulas in Section 2.4 and the estimates (2.4.8a–c) of α, β and γ, an approximate expression in which α, β and γ associated with each turning point have approximately disappeared, one obtains, by using the connection formulas (3.30.7a,b) but disregarding their one-directional nature, a result that is in general not correct. As examples of situations in which one obtains quite erroneous results by using a connection formula when the conditions for its validity are not fulfilled we refer the reader to (3.10.13), (3.13.4a,c), (3.19.12), (3.20.3), (3.20.7b) and (3.45.17). See also Fröman and Fröman (1998).

3.31 Quantization condition for a particle in a smooth potential well, limited on one side by an impenetrable wall and on the other side by a smooth, infinitely thick potential barrier, and in particular for a particle in a uniform gravitational field limited from below by an impenetrable plane surface

Derive the quantization condition for a non-relativistic quantal particle moving in a one-dimensional smooth potential well limited on one side by an impenetrable wall at $x = 0$ and on the other side by the turning point $t(> 0)$ delimiting a smooth, infinitely thick potential barrier; see Fig. 3.31.1.

Then calculate the energy levels of a quantal particle in a uniform gravitational field limited from below by an impenetrable plane surface, by using the quantization condition derived, as well as by using the Airy function.

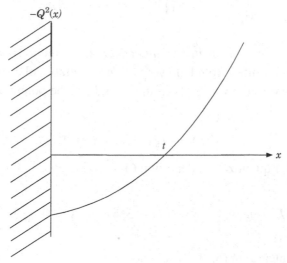

Figure 3.31.1 In the case of a uniform gravitational field, the plane surface is horizontal.

Solution. Since the wave function decreases exponentially as x recedes from t into the barrier, we can write

$$\psi(x) = D|q^{-1/2}(x)|\exp\left[-\left|\int_{(t)}^{x} q(z)dz\right|\right], \qquad x > t, \qquad (3.31.1)$$

and obtain by means of the connection formula (3.10.10′)

$$\psi(x) = 2D|q^{-1/2}(x)|\cos\left[\left|\int_{(t)}^{x} q(z)dz\right| - \pi/4\right], \qquad 0 < x < t. \quad (3.31.2)$$

In order for the wave function (3.31.2) to fulfil the boundary condition $\psi(0) = 0$, we have to put

$$\left|\int_{0}^{(t)} q(z)dz\right| = (s + 3/4)\pi, \qquad (3.31.3)$$

where s is a non-negative integer.

With the square of the base function chosen to be

$$Q^2(x) = R(x) = \frac{2m}{\hbar^2}[E - V(x)], \qquad (3.31.4)$$

the quantization condition (3.31.3) in the first-order approximation is

$$\int_{0}^{t} \left\{\frac{2m}{\hbar^2}[E - V(x)]\right\}^{1/2} dx = (s + 3/4)\pi. \qquad (3.31.5)$$

Particularization to a uniform gravitational field. For a quantal particle in a uniform gravitational field limited from below by an impenetrable plane surface at $x = 0$, which corresponds to the vertical wall in Fig. 3.31.1, the potential is

$$V(x) = +\infty \quad \text{for} \quad x \le 0, \qquad (3.31.6a)$$
$$V(x) = mgx \quad \text{for} \quad x > 0, \qquad (3.31.6b)$$

and the first-order quantization condition (3.31.5) is thus

$$\int_{0}^{E/(mg)} \left[\frac{2m}{\hbar^2}(E - mgx)\right]^{1/2} dx = (s + 3/4)\pi, \qquad s = 0, 1, 2, \ldots. \qquad (3.31.7)$$

Evaluating the integral in (3.31.7), we get

$$E = \frac{1}{2}(9\pi^2 mg^2\hbar^2)^{1/3}(s + 3/4)^{2/3}, \qquad s = 0, 1, 2, \ldots. \qquad (3.31.8)$$

The problem concerning the energy levels of a quantal particle in a uniform gravitational field limited from below by an impenetrable plane surface has also been treated by ter Haar (1964), Problem 1.27, who gives the correct result, and by Flügge (1974), Problem 119, whose solution is, however, not correct. Note that his formula (119.7) does not agree with his formulas (40.14) and (40.3).

Alternative solution for a uniform gravitational field with the use of the Airy function. We write for $x > 0$ the Schrödinger equation with the potential (3.31.6a,b), i.e.,

$$\frac{d^2\psi}{dx^2} + \frac{2m}{\hbar^2}(E - mgx)\psi = 0, \quad x > 0, \tag{3.31.9}$$

as

$$\frac{d^2\psi}{dz^2} - z\psi = 0, \tag{3.31.10a}$$

$$-A < z < +\infty, \tag{3.31.10b}$$

where

$$z = \left(\frac{2m^2 g}{\hbar^2}\right)^{1/3}\left(x - \frac{E}{mg}\right) = \left(\frac{2m^2 g}{\hbar^2}\right)^{1/3} x - A, \tag{3.31.11a}$$

$$A = 2E(4mg^2\hbar^2)^{-1/3} > 0. \tag{3.31.11b}$$

The wave function fulfilling the condition that ψ tends to zero as $x \to +\infty$, i.e., as $z \to +\infty$, is (except for an arbitrary constant factor) the Airy function $\psi = \text{Ai}(z)$. By requiring this wave function to be equal to zero for $x = 0$, i.e., for $z = -A$, we obtain the exact quantization condition

$$\text{Ai}(-A) = 0. \tag{3.31.12}$$

When A is sufficiently large, we have the asymptotic formula

$$\text{Ai}(-A) \sim \pi^{-1/2} A^{-1/4} \sin\left(\tfrac{2}{3} A^{3/2} + \pi/4\right), \tag{3.31.13}$$

and in this approximation we obtain from (3.31.12)

$$\tfrac{2}{3} A^{3/2} = (s + 3/4)\pi, \quad s = 0, 1, 2, \ldots. \tag{3.31.14}$$

Inserting the definition (3.31.11b) of A into (3.31.14), we obtain the first-order formula (3.31.8).

3.32 Energy spectrum of a non-relativistic particle in a potential proportional to $\cot^2(x/a_0)$, where $0 < x/a_0 < \pi$ and a_0 is a quantity with the dimension of length, e.g. the Bohr radius

Determine the energy levels of a quantal particle in the potential $V(x) = V_0 \cot^2(x/a_0)$ where $V_0 > 0$ and $0 < x/a_0 < \pi$. This is an example of a single-well potential with non-linear, very steep walls. It appears in connection with the free rotation of diatomic molecules and in the theory of Legendre functions; see Fröman, Fröman and Larsson (1994) and Larsson and Fröman (1994).

Solution. The Schrödinger equation can be written as

$$d^2\psi/dz^2 + R(z)\psi = 0, \tag{3.32.1a}$$

$$R(z) = A - B\cot^2 z, \tag{3.32.1b}$$

where

$$A = 2mEa_0^2/\hbar^2, \tag{3.32.2a}$$

$$B = 2mV_0a_0^2/\hbar^2, \tag{3.32.2b}$$

$$z = x/a_0. \tag{3.32.2c}$$

Recalling (2.2.12c), we choose the square of the base function to be

$$Q^2(z) = a - b\cot^2 z, \tag{3.32.3a}$$

$$b = B + 1/4 \,(>1/4), \tag{3.32.3b}$$

where the choice of b implies that the phase-integral approximation remains valid in the neighbourhood of the terminal points $z = 0$ and $z = \pi$ of the physically relevant interval for the variable z, and a is a so far unspecified constant which is assumed to differ not too much from A. The phase of the base function $Q(z)$ is chosen to be positive on the upper lip of the cut along the real axis between the turning points $z = \arctan[\pm(b/a)^{1/2}]$, $0 < \arctan < \pi$.

When the phase-integral approximation of the order $2N + 1$ is used, the quantization condition is given by (3.27.6a), (3.27.7a) and (3.27.7b'), i.e., by

$$\sum_{n=0}^{N} L^{(2n+1)} = (s + 1/2)\pi, \quad s = 0, 1, 2, \ldots, \tag{3.32.4a}$$

$$L^{(2n+1)} = \frac{1}{2}\int_{\Lambda} Z_{2n} Q(z)dz, \tag{3.32.4b}$$

with Z_{2n} given by (2.2.9a–d), (2.2.4) and (2.2.1), and with the integration performed in the negative sense along a contour Λ encircling the turning points, but no other transition point. Introducing in (3.32.4b) a new integration variable u by putting

$$u = \tan(z - \pi/2) = -\cot z, \tag{3.32.5}$$

and denoting by Λ_u the contour in the complex u-plane that corresponds to the contour Λ in the complex z-plane, we obtain with the aid of residue calculus

$$L^{(2n+1)} = \frac{1}{2} \int_{\Lambda_u} Z_{2n} Q \frac{du}{1 + u^2}$$

$$= \pi i \left(\operatorname*{Res}_{u=+i} \frac{Z_{2n} Q}{1 + u^2} + \operatorname*{Res}_{u=-i} \frac{Z_{2n} Q}{1 + u^2} \right) + \frac{1}{2} \int_{\Lambda'_u} Z_{2n} Q \frac{du}{1 + u^2}, \tag{3.32.6}$$

where Λ'_u is a contour encircling in the negative sense the part of the real u-axis corresponding to the classically allowed region, as well as the points $u = \pm i$, which lie outside the contour Λ_u. Noting that Q is approximately equal to $-ib^{1/2}u$ for large values of u and that Z_{2n}, which is given by (2.2.9a–d), tends to the Kronecker symbol $\delta_{n,0}$ as $|u| \to \infty$, we easily find that

$$L^{(2n+1)} = \frac{1}{2}\pi(a + b)^{1/2}[(Z_{2n})_{u=+i} + (Z_{2n})_{u=-i}] - \pi b^{1/2}\delta_{n,0}, \tag{3.32.7}$$

where $(a + b)^{1/2}$ and $b^{1/2}$ denote positive quantities.

To calculate expressions for the quantities $(Z_{2n})_{u=\pm i}$ we need formulas for $(\varepsilon_0)_{u=\pm i}$. Since $d(\cot z)/dz = -(1 + \cot^2 z) = -(1 + u^2) = 0$ when $u = \pm i$, the derivative $d^2 Q^{-1/2}/dz^2$ is equal to zero when $u = \pm i$. From the definition (2.2.1) of ε_0 along with (3.32.1b), (3.32.3a,b) and the definition (3.32.5) of u we therefore get

$$(\varepsilon_0)_{u=\pm i} = \left(\frac{R - Q^2}{Q^2} \right)_{u=\pm i} = \frac{A - 1/4 - a}{a + b}. \tag{3.32.8}$$

Using (2.2.9a–c) and (3.32.8), we obtain from (3.32.7)

$$L^{(1)} = \pi\left[(a + b)^{1/2} - b^{1/2}\right], \tag{3.32.9a}$$

$$L^{(3)} = \frac{1}{2}\pi(A - 1/4 - a)(a + b)^{-1/2}, \tag{3.32.9b}$$

$$L^{(5)} = -\frac{1}{8}\pi(A - 1/4 - a)^2(a + b)^{-3/2}. \tag{3.32.9c}$$

By choosing $a = A - 1/4$ we achieve the result that the higher-order contributions (3.32.9b,c) vanish and that thus the quantization condition (3.32.4a) in any order of approximation is

$$L^{(1)} = (s + 1/2)\pi, \quad s = 0, 1, 2, \ldots. \tag{3.32.10}$$

Inserting (3.32.9a) along with (3.32.3b) and $a = A - 1/4$ into (3.32.10), we obtain

$$A = [s + 1/2 + (B + 1/4)^{1/2}]^2 - B \qquad (3.32.11)$$

Inserting (3.32.2a,b) into (3.32.11), we obtain the formula

$$E = \{(s + 1/2)\hbar/(2ma_0^2)^{1/2} + [V_0 + \hbar^2/(8ma_0^2)]^{1/2}\}^2 - V_0, \quad (3.32.12)$$

which yields the energy eigenvalues exactly. We emphasize that the use of the phase-integral approximation generated from an unspecified base function is crucial in our treatment of this problem, and that the choice $a = A - 1/4$ that leads to the exact result (3.32.12) follows from the requirement that the higher-order contributions to the quantization condition vanish. This result could not have been obtained by means of the JWKB approximation, since there is no unspecified base function in that approximation.

The solution of this problem shows that for particular single-well potentials one can obtain the eigenvalues exactly from the phase-integral quantization condition by determining the base function such that the eigenvalues coincide in the first- and third-order approximations; such potentials have been considered by Rosenzweig and Krieger (1968) and Krieger (1969). For most potentials this is, however, not possible, but by choosing the base function such that the values of the energy coincide in the first- and third-order approximations, one obtains optimum results from the first-order quantization condition. One can thus make the first-order quantization condition as accurate as the more complicated third-order quantization condition, which may be an advantage in analytical work. It is not possible to achieve this result if one uses the JWKB approximation, since no a priori unspecified base function appears in that approximation. By ensuring that the above-mentioned exactness is fulfilled in the limit of a parameter value, for which the phase-integral results would not be accurate without this adaptation, one can also extend the region of validity of the phase-integral treatment; see p. 12 in Fröman, Fröman and Larsson (1994).

For a further discussion of quantization conditions associated with single-well potentials with very steep walls we refer to Fröman and Fröman (1978).

3.33 Determination of a *one-dimensional*, smooth, single-well potential from the energy spectrum of the bound states

The energy levels E_s of a *one-dimensional*, smooth, single-well potential $V(x)$ are assumed to be known as a function of the quantum number $s = 0, 1, 2, \ldots$. Investigate to what extent it is possible to determine $V(x)$ from this knowledge. Use the resulting formula to determine the potential which is mirror symmetric with respect to the origin and has the energy levels $E_s = (s + 1/2)\hbar\omega, s = 0, 1, 2, \ldots$.

Solution. One can formally consider the quantum number s as a continuous variable. The energy E is then known as a function of s, and one can determine s as a function $s(E)$ of E. From the first-order quantization condition

$$\left(\frac{2m}{\hbar^2}\right)^{1/2} \int_{t'(E)}^{t''(E)} [E - V(x)]^{1/2}\, dx = (s + 1/2)\pi \qquad (3.33.1)$$

one obtains

$$s(E) = \frac{1}{\pi} \left(\frac{2m}{\hbar^2}\right)^{1/2} \int_{t'(E)}^{t''(E)} [E - V(x)]^{1/2}\, dx - 1/2. \qquad (3.33.2)$$

The eigenvalues of the energy E correspond to non-negative integer values for the function $s(E)$. The minimum value V_0 of the potential $V(x)$ is according to (3.33.2) obtained from the equation

$$s(V_0) = -1/2. \qquad (3.33.3)$$

Differentiating (3.33.2) with respect to E, and replacing E by E', we obtain

$$\frac{ds(E')}{dE'} = \frac{1}{2\pi} \left(\frac{2m}{\hbar^2}\right)^{1/2} \int_{t'(E')}^{t''(E')} \frac{dx}{[E' - V(x)]^{1/2}}. \qquad (3.33.4)$$

Multiplying this equation by $(E - E')^{-1/2}$, and integrating over E' from V_0 to E, we get with the use of (3.33.4)

$$\int_{V_0}^{E} \frac{ds(E')/dE'}{(E - E')^{1/2}} dE' = \frac{1}{2\pi} \left(\frac{2m}{\hbar^2}\right)^{1/2} \int_{V_0}^{E} dE' \int_{t'(E')}^{t''(E')} \frac{dx}{\{(E - E')[E' - V(x)]\}^{1/2}}$$

$$= \frac{1}{2\pi} \left(\frac{2m}{\hbar^2}\right)^{1/2} \int_{t'(E)}^{t''(E)} dx \int_{V(x)}^{E} \frac{dE'}{\{(E - E')[E' - V(x)]\}^{1/2}}.$$

$$(3.33.5)$$

Since the integration here with respect to E' yields π, we obtain

$$\int_{V_0}^{E} \frac{ds(E')/dE'}{(E - E')^{1/2}} dE' = \frac{1}{2} \left(\frac{2m}{\hbar^2}\right)^{1/2} \int_{t'(E)}^{t''(E)} dx = \frac{1}{2} \left(\frac{2m}{\hbar^2}\right)^{1/2} [t''(E) - t'(E)],$$

$$(3.33.6)$$

i.e.,

$$t''(E) - t'(E) = 2\left(\frac{\hbar^2}{2m}\right)^{1/2} \int_{V_0}^{E} \frac{ds(E')/dE'}{(E - E')^{1/2}} dE'. \qquad (3.33.7)$$

This formula shows that from knowledge of the energy levels of a *one-dimensional* single-well potential (which depend only on the single quantum number s) one can only calculate the difference $t''(E) - t'(E)$. Because of this fact it is possible (in the first-order approximation) to construct an infinite number of potentials with the given energy spectrum. However, if we require that $V(x)$ be mirror symmetric with respect to the origin, we have $t'(E) = -t''(E)$, and hence (3.33.7) can be written as

$$-t'(E) = t''(E) = \left(\frac{\hbar^2}{2m}\right)^{1/2} \int_{V_0}^{E} \frac{ds(E')/dE'}{[V(x) - E']^{1/2}} dE'. \qquad (3.33.8)$$

Letting x denote either $t'(E)$ or $t''(E)$, we have $V(x) = E$, and we obtain from (3.33.8) the formula

$$x = \pm\left(\frac{\hbar^2}{2m}\right)^{1/2} \int_{V_0}^{V(x)} \frac{ds(E')/dE'}{[V(x) - E']^{1/2}} dE', \qquad (3.33.9)$$

from which the potential $V(x)$ can be determined.

If the energy levels are $E_s = (s + 1/2)\hbar\omega, s = 0, 1, 2, \ldots$, we have $s(E) = E/(\hbar\omega) - 1/2$, and from (3.33.3) we obtain $V_0 = 0$. For the integral in (3.33.9) we therefore get

$$\int_{V_0}^{V(x)} \frac{ds(E')/dE'}{[V(x) - E']^{1/2}} dE' = \int_{0}^{V(x)} \frac{1/(\hbar\omega)}{[V(x) - E']^{1/2}} dE' = \frac{2}{\hbar\omega} [V(x)]^{1/2}, \qquad (3.33.10)$$

which, when inserted into (3.33.9), gives $x = \pm[2V(x)/(m\omega^2)]^{1/2}$, i.e., the harmonic oscillator potential $V(x) = m\omega^2 x^2/2$.

3.34 Determination of a *radial*, smooth, single-well potential from the energy spectrum of the bound states

A non-relativistic quantal particle moving in the *radial* single-well potential $V(r)$ has the bound-state energy E, which is assumed to be a known function of the angular momentum quantum number $l (= 0, 1, 2, \ldots)$ and the radial quantum number $s (= 0, 1, 2, \ldots)$. Determine the potential $V(r)$.

Solution. The radial Schrödinger equation is

$$\frac{d^2\psi}{dr^2} + \frac{2m}{\hbar^2}\left[E - V(r) - \frac{\hbar^2 l(l+1)}{2mr^2}\right]\psi = 0, \tag{3.34.1}$$

and the first-order quantization condition is

$$L(l, E) = (s + 1/2)\pi, \quad s = 0, 1, 2, \ldots, \tag{3.34.2}$$

where, with the base function chosen according to (2.2.12a),

$$L(l, E) = \left(\frac{2m}{\hbar^2}\right)^{1/2} \int\limits_{t'(l,E)}^{t''(l,E)} [E - \bar{V}(r)]^{1/2} dr, \tag{3.34.3a}$$

$$\bar{V}(r) = V(r) + \frac{\hbar^2(l + 1/2)^2}{2mr^2}, \tag{3.34.3b}$$

$t'(l, E)$ and $t''(l, E)$ being the two turning points, i.e., the two real zeros of $E - \bar{V}(r)$. The quantum numbers l and s, which for non-negative integer values determine the energy eigenvalues, are formally considered as continuous variables. When the bound-state energy E is known as a function of l and s, one can solve (3.34.2) with respect to s, thus obtaining the function $s(l, E)$:

$$s(l, E) = L(l, E)/\pi - 1/2. \tag{3.34.4}$$

We shall now show how one can determine $\bar{V}(r)$ and hence $V(r)$ from the function $s(l, E)$.

The 'inclusion' $I(l, E)$ for the potential well, which is defined by

$$I(l, E) = \int\limits_{t'(l,E)}^{t''(l,E)} [E - \bar{V}(r)]dr, \tag{3.34.5}$$

increases as E increases, but decreases as l increases. From this definition and the definition (3.34.3b) of $\bar{V}(r)$ it follows that

$$\frac{\partial I(l, E)}{\partial E} = \int\limits_{t'(l,E)}^{t''(l,E)} dr = t''(l, E) - t'(l, E), \tag{3.34.6a}$$

$$\frac{\partial I(l, E)}{\partial l} = -\frac{(l + 1/2)\hbar^2}{m}\int\limits_{t'(l,E)}^{t''(l,E)}\frac{dr}{r^2} = -\frac{(l + 1/2)\hbar^2}{m}\frac{t''(l, E) - t'(l, E)}{t'(l, E)t''(l, E)}. \tag{3.34.6b}$$

Using the fact that in a radial problem $t'(l, E)$ and $t''(l, E)$ $(> t'(l, E))$ are positive, we obtain from (3.34.6a,b)

$$t'(l, E) = \left[\left(\frac{1}{2}\frac{\partial I(l, E)}{\partial E}\right)^2 + \frac{(2l + 1)\hbar^2}{m}\frac{\frac{1}{2}\partial I(l, E)/\partial E}{-\partial I(l, E)/\partial l}\right]^{1/2} - \frac{1}{2}\frac{\partial I(l, E)}{\partial E},$$

(3.34.7a)

$$t''(l, E) = \left[\left(\frac{1}{2}\frac{\partial I(l, E)}{\partial E}\right)^2 + \frac{(2l + 1)\hbar^2}{m}\frac{\frac{1}{2}\partial I(l, E)/\partial E}{-\partial I(l, E)/\partial l}\right]^{1/2} + \frac{1}{2}\frac{\partial I(l, E)}{\partial E}.$$

(3.34.7b)

The derivatives $\partial I(l, E)/\partial l$ and $\partial I(l, E)/\partial E$ in (3.34.7a,b) can be expressed in terms of the derivatives of $L(l,E)$, and hence in terms of the derivatives of $s(l, E)$, with respect to l and E. For this purpose one uses the identity

$$E - \bar{V}(r) = \frac{2}{\pi}\int_{\bar{V}(r)}^{E}\left[\frac{E' - \bar{V}(r)}{E - E'}\right]^{1/2}dE',$$

(3.34.8)

which one can easily verify by evaluating the integral on the right-hand side. Inserting this expression for $E - \bar{V}(r)$ into the definition (3.34.5) of $I(l,E)$, reversing the orders of integration, and using the definition (3.34.3a) of $L(l,E)$, we get the formula

$$I(l, E) = \frac{2}{\pi}\left(\frac{\hbar^2}{2m}\right)^{1/2}\int_{\bar{V}_0}^{E}\frac{L(l, E')}{(E - E')^{1/2}}dE',$$

(3.34.9)

where $\bar{V}_0$ denotes the minimum value of $\bar{V}(r)$. From the definition (3.34.3a) of $L(l,E)$ it follows that one can obtain $\bar{V}_0$ by solving the equation

$$L(l, \bar{V}_0) = 0,$$

(3.34.10)

which because of (3.34.4) can be written as

$$s(l, \bar{V}_0) = -1/2.$$

(3.34.11)

By means of partial integration and use of (3.34.10) we can write (3.34.9) in the alternative form

$$I(l, E) = \frac{4}{\pi}\left(\frac{\hbar^2}{2m}\right)^{1/2}\int_{\bar{V}_0}^{E}\frac{\partial L(l, E')}{\partial E'}(E - E')^{1/2}dE'.$$

(3.34.12)

We differentiate (3.34.9) with respect to l and (3.34.12) with respect to E, getting

$$-\frac{\partial I(l, E)}{\partial l} = \frac{2}{\pi} \left(\frac{\hbar^2}{2m}\right)^{1/2} \int_{\bar{V}_0}^{E} \frac{-\partial L(l, E')/\partial l}{(E - E')^{1/2}} dE', \qquad (3.34.13a)$$

$$\frac{1}{2} \frac{\partial I(l, E)}{\partial E} = \frac{1}{\pi} \left(\frac{\hbar^2}{2m}\right)^{1/2} \int_{\bar{V}_0}^{E} \frac{\partial L(l, E')/\partial E'}{(E - E')^{1/2}} dE'. \qquad (3.34.13b)$$

Expressing $L(l, E')$ in terms of $s(l, E')$ using (3.34.4), we obtain

$$-\frac{\partial I(l, E)}{\partial l} = 2 \left(\frac{\hbar^2}{2m}\right)^{1/2} \int_{\bar{V}_0}^{E} \frac{-\partial s(l, E')/\partial l}{(E - E')^{1/2}} dE', \qquad (3.34.14a)$$

$$\frac{1}{2} \frac{\partial I(l, E)}{\partial E} = \left(\frac{\hbar^2}{2m}\right)^{1/2} \int_{\bar{V}_0}^{E} \frac{\partial s(l, E')/\partial E'}{(E - E')^{1/2}} dE'. \qquad (3.34.14b)$$

Using these formulas in (3.34.7a,b) we can obtain explicit expressions for the functions $t'(l, E)$ and $t''(l, E)$, which determine $\bar{V}(r)$. The physical potential $V(r)$ is then obtained from (3.34.3b). The procedure described is a version of the well-known Rydberg–Klein–Rees procedure for determining a radial single-well potential from the knowledge of the energy levels of the bound states.

In an analogous way, Wheeler (1976) and Cole and Good (1978) have shown how the shape of a potential barrier can be determined from knowledge of the transmission coefficient as a function of the energy, and Fröman and Fröman (1989) have shown how the energies and widths of shape resonances, i.e., quasi-stationary levels for a quantal particle in a spherically symmetric potential, can be used to determine the potential.

In Section 3.35 we shall use formulas from the present problem to determine the physical potential corresponding to a particular energy spectrum.

3.35 Determination of the *radial*, single-well potential, when the energy eigenvalues are $-mZ^2e^4/[2\hbar^2(l + s + 1)^2]$, where l is the angular momentum quantum number, and s is the radial quantum number

A non-relativistic quantal particle moving in a spherically symmetric potential $V(r)$ has the bound-state energy levels

$$E = -mZ^2e^4/[2\hbar^2(l + s + 1)^2], \qquad (3.35.1)$$

where $l(= 0, 1, 2, \ldots)$ is the angular momentum quantum number, and $s (= 0, 1, 2, \ldots)$ is the radial quantum number. Use the formulas derived in Section 3.34 to determine $V(r)$.

Solution. Solving (3.35.1) with respect to s, we obtain for the function $s(l, E)$ the expression

$$s(l, E) = \frac{Ze^2}{\hbar} \left(\frac{m}{-2E} \right)^{1/2} - l - 1. \tag{3.35.2}$$

With this expression for $s(l, E)$ we obtain from (3.34.11)

$$\bar{V}_0 = -\frac{mZ^2e^4}{2(l + 1/2)^2\hbar^2}. \tag{3.35.3}$$

Partial differentiation of expression (3.35.2) for $s(l, E)$ with respect to l and s yields

$$-\frac{\partial s(l, E)}{\partial l} = 1, \tag{3.35.4a}$$

$$\frac{\partial s(l, E)}{\partial E} = \frac{m^{1/2}Ze^2}{\hbar(-2E)^{3/2}}. \tag{3.35.4b}$$

For the integrals in (3.34.14a,b) we obtain by using (3.35.4a,b) and (3.35.3) the formulas

$$\int_{\bar{V}_0}^{E} \frac{-\partial s(l, E')/\partial l}{(E - E')^{1/2}} dE' = \int_{\bar{V}_0}^{E} \frac{dE'}{(E - E')^{1/2}} = 2(E - \bar{V}_0)^{1/2} = 2\left[\frac{mZ^2e^4}{2(l + 1/2)^2\hbar^2} + E \right]^{1/2},$$

$$\tag{3.35.5a}$$

$$\int_{\bar{V}_0}^{E} \frac{-\partial s(l, E')/\partial E'}{(E - E')^{1/2}} dE' = \frac{m^{1/2}Ze^2}{\hbar} \int_{\bar{V}_0}^{E} \frac{dE'}{(-2E')^{3/2}(E - E')^{1/2}}$$

$$= \frac{m^{1/2}Ze^2}{\hbar E} \int_{\bar{V}_0}^{E} \frac{\partial}{\partial E'} \left(\frac{1}{2} - \frac{E}{2E'} \right)^{1/2} dE'$$

$$= -\frac{m^{1/2}Ze^2}{\hbar E} \left(\frac{1}{2} - \frac{E}{2\bar{V}_0} \right)^{1/2}$$

$$= -\frac{m^{1/2}Ze^2}{\hbar E} \left[\frac{1}{2} + \frac{(l + 1/2)^2\hbar^2 E}{mZ^2e^4} \right]^{1/2}$$

$$= \frac{l + 1/2}{-E} \left[\frac{mZ^2e^4}{2(l + 1/2)^2\hbar^2} + E \right]^{1/2}. \tag{3.35.5b}$$

Inserting (3.35.5a) into (3.34.14a) and (3.35.5b) into (3.34.14b), we obtain

$$
-\frac{\partial I(l, E)}{\partial l} = 4\left(\frac{\hbar^2}{2m}\right)^{1/2}\left[\frac{mZ^2e^4}{2(l + 1/2)^2\hbar^2} + E\right]^{1/2}, \qquad (3.35.6a)
$$

$$
\frac{1}{2}\frac{\partial I(l, E)}{\partial E} = \left(\frac{\hbar^2}{2m}\right)^{1/2}\frac{l + 1/2}{-E}\left[\frac{mZ^2e^4}{2(l + 1/2)^2\hbar^2} + E\right]^{1/2}, \quad (3.35.6b)
$$

and hence (since $E < 0$)

$$
\left\{\left[\frac{1}{2}\frac{\partial I(l, E)}{\partial E}\right]^2 + \frac{(2l + 1)\hbar^2}{m}\frac{\frac{1}{2}\partial I(l, E)/\partial E}{-\partial I(l, E)/\partial l}\right\}^{1/2}
$$

$$
= \left\{\frac{\hbar^2}{2m}\left(\frac{l + 1/2}{-E}\right)^2\left[\frac{mZ^2e^4}{2(l + 1/2)^2\hbar^2} + E\right] + \frac{(2l + 1)\hbar^2}{m}\frac{l + 1/2}{-4E}\right\}^{1/2} = \frac{Ze^2}{-2E}.
$$

$$
(3.35.7)
$$

Inserting (3.35.6b) and (3.35.7) into the formulas (3.34.7a,b) for $t'(l, E)$ and $t''(l, E)$ [$> t'(l, E)$], we obtain

$$
t'(l, E) = \frac{Ze^2}{-2E} - \left(\frac{\hbar^2}{2m}\right)^{1/2}\frac{l + 1/2}{-E}\left[\frac{mZ^2e^4}{2(l + 1/2)^2\hbar^2} + E\right]^{1/2}, \quad (3.35.8a)
$$

$$
t''(l, E) = \frac{Ze^2}{-2E} + \left(\frac{\hbar^2}{2m}\right)^{1/2}\frac{l + 1/2}{-E}\left[\frac{mZ^2e^4}{2(l + 1/2)^2\hbar^2} + E\right]^{1/2}. \quad (3.35.8b)
$$

Letting r denote either $t'(l, E)$ or $t''(l, E)$, we have $\bar{V}(r) = E$, and hence we can write (3.35.8a,b) as

$$
r = \frac{Ze^2}{-2\bar{V}(r)} \pm \left(\frac{\hbar^2}{2m}\right)^{1/2}\frac{l + 1/2}{-\bar{V}(r)}\left[\frac{mZ^2e^4}{2(l + 1/2)^2\hbar^2} + \bar{V}(r)\right]^{1/2}. \quad (3.35.9)
$$

From (3.35.9) we obtain the formula

$$
\left[r + \frac{Ze^2}{2\bar{V}(r)}\right]^2 = \frac{\hbar^2}{2m}\left[\frac{l + 1/2}{-\bar{V}(r)}\right]^2\left[\frac{mZ^2e^4}{2(l + 1/2)^2\hbar^2} + \bar{V}(r)\right], \quad (3.35.10)
$$

i.e.,

$$
\bar{V}(r)\left[\bar{V}(r) + \frac{Ze^2}{r} - \frac{(l + 1/2)^2\hbar^2}{2mr^2}\right] = 0, \qquad (3.35.11)
$$

and hence, since $\bar{V}(r)$ is not identically equal to zero,

$$\bar{V}(r) = -\frac{Ze^2}{r} + \frac{\hbar^2(l+1/2)^2}{2mr^2} \qquad (3.35.12)$$

and thus, according (3.34.3b),

$$V(r) = -Ze^2/r. \qquad (3.35.13)$$

This is the potential of a hydrogenic atom or ion.

3.36 Exact formula for the normalization integral for the wave function pertaining to a bound state of a particle in a radial potential

Consider the radial, real wave function $u(r, E_s)$ for a bound state of a quantal particle with the radial quantum number s and the energy E_s in a radial single-well potential and derive a general, exact formula for the normalization integral $\int_0^\infty u^2(r, E_s)dr$.

Solution. With obvious notation the radial Schrödinger equation is

$$d^2u/dr^2 + R(r, E)u = 0, \qquad (3.36.1a)$$
$$R(r, E) = 2m[E - V(r)]/\hbar^2 - l(l+1)/r^2, \qquad (3.36.1b)$$

and the boundary conditions for the problem are

$$u(0, E_s) = 0, \qquad (3.36.2a)$$
$$u(+\infty, E_s) = 0. \qquad (3.36.2b)$$

For any value of E, which thus need not correspond to a bound state, we let $u_1(r, E)$ and $u_2(r, E)$ be real solutions of (3.36.1a) that for all values of E vanish at $r = 0$ and $r = +\infty$, respectively, i.e., fulfil the boundary conditions

$$u_1(0, E) = 0, \qquad (3.36.3a)$$
$$u_2(+\infty, E) = 0 \qquad (3.36.3b)$$

and are equal to $u(r, E_s)$ when $E = E_s$, i.e., fulfil the conditions

$$u_1(r, E_s) \equiv u(r, E_s), \qquad (3.36.4a)$$
$$u_2(r, E_s) \equiv u(r, E_s). \qquad (3.36.4b)$$

From (3.36.3a,b) it follows that

$$\frac{\partial u_1(0, E)}{\partial E} = 0, \qquad (3.36.5a)$$
$$\frac{\partial u_2(+\infty, E)}{\partial E} = 0. \qquad (3.36.5b)$$

When E is not equal to an eigenvalue E_s, the solutions $u_1(r, E)$ and $u_2(r, E)$ are linearly independent and do not tend to zero as r tends to infinity and to zero, respectively.

Denoting the derivative with respect to r by a prime, we obtain using (3.36.1a,b)

$$\frac{d}{dr}\left(u'_k\frac{\partial u_k}{\partial E} - u_k\frac{\partial u'_k}{\partial E}\right) = u''_k\frac{\partial u_k}{\partial E} - u_k\frac{\partial u''_k}{\partial E} = -Ru_k\frac{\partial u_k}{\partial E} + u_k\frac{\partial}{\partial E}(Ru_k)$$

$$= \frac{\partial R}{\partial E}u_k^2 = \frac{2m}{\hbar^2}u_k^2, \quad k = 1, 2, \tag{3.36.6}$$

i.e.,

$$u_1^2 = \frac{\hbar^2}{2m}\frac{d}{dr}\left(u'_1\frac{\partial u_1}{\partial E} - u_1\frac{\partial u'_1}{\partial E}\right), \tag{3.36.7a}$$

$$u_2^2 = \frac{\hbar^2}{2m}\frac{d}{dr}\left(u'_2\frac{\partial u_2}{\partial E} - u_2\frac{\partial u'_2}{\partial E}\right). \tag{3.36.7b}$$

Because of (3.36.3a,b) and (3.36.5a,b) the functions to the right of d/dr in (3.36.7a) and (3.36.4b), respectively, are equal to zero for $r = 0$ and $r = +\infty$, respectively, and therefore we obtain from (3.36.7a,b)

$$\int_0^r u_1^2 dr = \frac{\hbar^2}{2m}\left(u'_1\frac{\partial u_1}{\partial E} - u_1\frac{\partial u'_1}{\partial E}\right), \tag{3.36.8a}$$

$$\int_r^\infty u_2^2 dr = -\frac{\hbar^2}{2m}\left(u'_2\frac{\partial u_2}{\partial E} - u_2\frac{\partial u'_2}{\partial E}\right). \tag{3.36.8b}$$

Using (3.36.4a,b) we obtain from (3.36.8a,b)

$$\int_0^r u^2(r, E_s)dr = \int_0^r u_1^2(r, E_s)dr = \frac{\hbar^2}{2m}\left(u'_1\frac{\partial u_1}{\partial E} - u_1\frac{\partial u'_1}{\partial E}\right)_{E=E_s}$$

$$= \frac{\hbar^2}{2m}\left(u'_2\frac{\partial u_1}{\partial E} - u_2\frac{\partial u'_1}{\partial E}\right)_{E=E_s}, \tag{3.36.9a}$$

$$\int_r^\infty u^2(r, E_s)dr = \int_r^\infty u_2^2(r, E_s)dr = -\frac{\hbar^2}{2m}\left(u'_2\frac{\partial u_2}{\partial E} - u_2\frac{\partial u'_2}{\partial E}\right)_{E=E_s}$$

$$= -\frac{\hbar^2}{2m}\left(u'_1\frac{\partial u_2}{\partial E} - u_1\frac{\partial u'_2}{\partial E}\right)_{E=E_s}. \tag{3.36.9b}$$

Adding (3.36.9a) and (3.36.9b), we obtain

$$\int_0^\infty u^2(r, E_s)dr = \frac{\hbar^2}{2m}\left(u_2'\frac{\partial u_1}{\partial E} - u_2\frac{\partial u_1'}{\partial E} - u_1'\frac{\partial u_2}{\partial E} + u_1\frac{\partial u_2'}{\partial E}\right)_{E=E_s}$$

$$= \frac{\hbar^2}{2m}\left(\frac{\partial}{\partial E}(u_1u_2' - u_2u_1')\right)_{E=E_s}, \qquad (3.36.10)$$

i.e.,

$$\int_0^\infty u^2(r, E_s)dr = \frac{\hbar^2}{2m}\left(\frac{dW}{dE}\right)_{E=E_s}, \qquad (3.36.11)$$

where W is the Wronskian

$$W = u_1u_2' - u_2u_1'. \qquad (3.36.12)$$

The normalization integral is thus related in a very simple way to the energy derivative of the r-independent Wronskian (3.36.12). This result was obtained by Yngve (1972), although with the limits of integration $-\infty$ and $+\infty$.

For the background of this problem and for applications of the resulting formulas we refer the reader to Yngve (1972, 1986, 1988), Fröman and Fröman (1972, 1981), P. O. Fröman (1974), N. Fröman (1978a,b, 1980), Thidé (1980), Linnæus and Düring (1985) and Fröman, Yngve and Fröman (1987).

3.37 Phase-integral formula for the normalized radial wave function pertaining to a bound state of a particle in a radial single-well potential

Use the exact formula given by (3.36.11) and (3.36.12) to derive the phase-integral formula in question.

Solution. The radial Schrödinger equation is given by (3.36.1a,b). We choose the base function to be

$$Q(r, E) = \{2m[E - V(r)]/\hbar^2 - (l + 1/2)^2/r^2\}^{1/2}. \qquad (3.37.1)$$

Denoting the turning points by t' and $t''(>t')$, and assuming that

$$\left|\int_{t'}^r q(r, E)dr\right| \to \infty \quad \text{as } r \to 0, \qquad (3.37.2a)$$

$$\left|\int_{t''}^r q(r, E)dr\right| \to \infty \quad \text{as } r \to +\infty, \qquad (3.37.2b)$$

we have for $u_1(r, E)$ and $u_2(r, E)$ in the classically forbidden regions $r < t'$ and $r > t''$, respectively, the phase-integral formulas

$$u_1(r, E) = D_1(E)|q^{-1/2}(r, E)|\exp\left[-\left|\int_{(t')}^{r} q(r, E)dr\right|\right], \quad r < t', \quad (3.37.3a)$$

$$u_2(r, E) = D_2(E)|q^{-1/2}(r, E)|\exp\left[-\left|\int_{(t'')}^{r} q(r, E)dr\right|\right], \quad r > t'', \quad (3.37.3b)$$

where $D_1(E)$ and $D_2(E)$ do not depend on r. Because of (3.37.2a,b) the conditions (3.36.2a,b) are fulfilled. The base function $Q(r, E)$ is chosen to be positive in the classically allowed region, i.e., for $t' < r < t''$. Therefore we obtain by means of the connection formula (3.10.10′) for tracing a solution from a classically forbidden region to an adjacent classically allowed region

$$u_1(r, E) = 2D_1(E)q^{-1/2}(r, E)\cos\left[\int_{(t')}^{r} q(r, E)dr - \pi/4\right], \quad t' < r < t'',$$

$$(3.37.4a)$$

$$u_2(r, E) = 2D_2(E)q^{-1/2}(r, E)\cos\left[\int_{r}^{(t'')} q(r, E)dr - \pi/4\right]$$

$$= 2D_2(E)q^{-1/2}(r, E)\cos\left[\int_{(t')}^{r} q(r, E)dr - L(E) + \pi/4\right],$$

$$t' < r < t'', \quad (3.37.4b)$$

where

$$L(E) = \int_{(t')}^{(t'')} q(r, E)dr = \frac{1}{2}\int_{\Lambda} q(r, E)dr. \quad (3.37.5)$$

Here Λ is a closed contour of integation that encircles the turning points t' and t'' in the direction for which the integral is positive. Imposing the conditions (3.36.4a,b), which imply that $u_1(r, E_s) = u_2(r, E_s)$, we obtain from (3.37.4a,b)

$$L(E_s) = (s + 1/2)\pi, \quad (3.37.6a)$$

$$D_2(E_s) = (-1)^s D_1(E_s), \quad (3.37.6b)$$

where s is a non-negative integer. Using (3.36.12) and (3.37.4a,b), we obtain

$$W = u_1 u_2' - u_2 u_1' = 4D_1(E)D_2(E)\sin[L(E) - \pi/2] \qquad (3.37.7)$$

and hence, using (3.37.6a,b),

$$\left(\frac{dW}{dE}\right)_{E=E_s} = \left[4D_1 D_2 \cos(L - \pi/2)\frac{dL}{dE}\right]_{E=E_s} = \left[4D_1^2\, dL/dE\right]_{E=E_s}. \qquad (3.37.8)$$

Inserting (3.37.8) into (3.36.11), we obtain

$$\int_0^\infty u^2(r, E_s)dr = \frac{2\hbar^2 D_1^2(E_s)}{m}\left(\frac{dL}{dE}\right)_{E=E_s}. \qquad (3.37.9)$$

Requiring that the integral on the left-hand side of (3.37.9) be equal to unity, we obtain

$$|D_1(E_s)| = \left[\frac{m/(2\hbar^2)}{dL/dE}\right]_{E=E_s}^{1/2}. \qquad (3.37.10)$$

Under the assumption that $R(r, E) - Q^2(r, E)$ is independent of E, N. Fröman (1974) and P. O. Fröman (1974) proved the approximate, but in general very accurate, formula

$$\frac{d}{dE}\int_\Lambda q(r, E)dr = \int_\Lambda \frac{1}{2}\frac{\partial R(r, E)}{\partial E}\frac{dr}{q(r, E)}, \qquad (3.37.11)$$

where on the right-hand side the contour Λ must for higher-order approximations encircle not only the turning points t' and t'', but also the associated zeros of $q(r, E_s)$ mentioned below (2.2.11). One can easily verify (3.37.11) for the first-order phase-integral approximation. There exists an analogous formula for the derivative of the phase integral with respect to an arbitrary parameter in the phase integrand; see (3.40.11). From (3.37.5), (3.37.11) and (3.36.1b) we obtain

$$\frac{dL}{dE} = \frac{m}{2\hbar^2}\int_\Lambda \frac{dr}{q(r, E)}, \qquad (3.37.12)$$

and therefore (3.37.10) can be written as

$$|D_1(E_s)| = \frac{1}{\left[\displaystyle\int_\Lambda dr/q(r, E_s)\right]^{1/2}}. \qquad (3.37.13)$$

In the dashed band in Fig. 3.28.1(c) and on the real axis to the left and to the right of the band the wave function is given according to (3.37.3a) and (3.37.13) by the approximate formula (cf. eqs. (30a,c) in P. O. Fröman (1974))

$$\psi(r, E_s) = \frac{\exp(i\pi/4)q^{-1/2}(r, E_s)\exp[-iw(r, E_s)]}{\left[\displaystyle\int_{\Lambda} dr/q(r, E_s)\right]^{1/2}}, \qquad (3.37.14a)$$

$$w(r, E_s) = \int_{(t')}^{r} q(r, E_s)dr, \qquad (3.37.14b)$$

if $q(r, E_s)$ is positive on the upper lip of a cut between t' and t'' and if we choose $D_1(E_s)$ to be positive. Using the first-order approximation, we obtain from (3.37.13)

$$D_1(E_s) = \frac{1}{\left[2\displaystyle\int_{t'}^{t''} dr/Q(r)\right]^{1/2}} = \left[\frac{m/\hbar}{2\displaystyle\int_{t'}^{t''} \frac{dr}{\hbar Q(r)/m}}\right]^{1/2} = \left(\frac{m/\hbar}{\tau}\right)^{1/2} = \left(\frac{mv}{\hbar}\right)^{1/2},$$

$$(3.37.15)$$

where τ is the time for a complete oscillation of a classical particle in the potential $V(r) + \hbar^2(l + 1/2)^2/(2mr^2)$, and $\nu = 1/\tau$ is the corresponding frequency; see Section 3.26.

3.38 Radial wave function $\psi(z)$ for an s-electron in a classically allowed region containing the origin, when the potential near the origin is dominated by a strong, attractive Coulomb singularity, and the normalization factor is chosen such that, when the radial variable z is dimensionless, $\psi(z)/z$ tends to unity as z tends to zero

If, for an s-electron, one chooses $Q(z) = R^{1/2}(z)$, the phase-integral approximation for the radial wave function breaks down close to the origin when the potential has a Coulomb singularity there. For the case in which the potential in the neighbourhood of the origin is dominated by an attractive Coulomb singularity at the origin use the comparison equation technique to derive, in the classically allowed region containing the origin, but sufficiently far away from it, a first-order phase-integral formula with $Q(z) = R^{1/2}(z)$ for the radial s-electron wave function $\psi(z)$ that is normalized such that $\psi(z)/z$ tends to unity as z tends to zero; here z is a dimensionless radial variable.

Solution. The radial Schrödinger equation for an s-electron, i.e., an electron with the angular momentum quantum number l equal to zero, is

$$d^2u/dr^2 + \frac{2m}{\hbar^2}[E - V(r)]u = 0, \tag{3.38.1}$$

where

$$V(r) = -Ze^2/r + \text{a function that is regular at } r = 0, \tag{3.38.2}$$

Z being assumed to be positive and sufficiently large. We introduce instead of r the dimensionless variable z and instead of $u(r)$ the dimensionsless wave function $\psi(z)$ by putting

$$z = r/a_0, \tag{3.38.3a}$$

$$\psi(z) = a_0^{1/2}u(r), \tag{3.38.3b}$$

where a_0 is a quantity of the dimension of length, e.g. the Bohr radius. The factor $a_0^{1/2}$ in (3.38.3b) makes the normalization integral the same for $\psi(z)$ as for $u(r)$. Using (3.38.3a,b) and (3.38.2), we write (3.38.1) as

$$\frac{d^2\psi}{dz^2} + R(z)\psi = 0, \tag{3.38.4a}$$

$$R(z) = \frac{2m\,a_0^2}{\hbar^2}[E - V(r)] = \frac{2mZe^2a_0}{\hbar^2 z} + \text{a function that is regular at } z = 0.$$

$$\tag{3.38.4b}$$

To treat (3.38.4a,b) by the comparison equation technique we put (cf. eq. (2.2.3) in Fröman and Fröman (1996))

$$\psi = A(d\rho/dz)^{-1/2}F_0(\eta, \rho), \tag{3.38.5}$$

where A is a constant normalization factor, ρ is a so far unknown function of z, and $F_0(\eta, \rho)$ is the regular Coulomb wave function that satisfies the differential equation

$$d^2F_0/d\rho^2 + (1 - 2\eta/\rho)F_0 = 0. \tag{3.38.6}$$

Since Z is assumed to be positive and sufficiently large, η is chosen to be negative and of a sufficiently large absolute value. This is the only information we need concerning η in the present problem. The variable ρ can be determined as a function of z from the formula (cf. eq. (2.2.20) in Fröman and Fröman (1996))

$$\int_0^{\rho} (1 - 2\eta/\rho)^{1/2} d\rho = \int_0^z Q(z)dz, \qquad (3.38.7)$$

where (cf. (3.38.4b))

$$Q(z) = R^{1/2}(z) = \{2ma_0^2 [E - V(r)]/\hbar^2\}^{1/2}$$

$$= \left(\frac{2mZe^2a_0}{\hbar^2 z} + \text{a function that is regular at } z = 0\right)^{1/2}. \quad (3.38.8)$$

In passing we remark that the condition

$$\int_0^{2\eta} (1 - 2\eta/\rho)^{1/2} d\rho = \int_0^t Q(z)dz, \qquad (3.38.9)$$

where t is the transition zero of $Q^2(z)$ corresponding to $\rho = 2\eta$, can be used for the determination of η as a function of t in cases where such information is needed. By evaluating the integrals in (3.38.7) for sufficiently small values of ρ and z, when $Q(z)$ is given by (3.38.8), we obtain approximately

$$\int_0^{\rho} (1 - 2\eta/\rho)^{1/2} d\rho \approx \int_0^{\rho} (-2\eta/\rho)^{1/2} d\rho = 2(-2\eta\rho)^{1/2}, \quad (3.38.10a)$$

$$\int_0^z Q(z)dz \approx \int_0^z [2mZe^2a_0/(\hbar^2 z)]^{1/2} dz = 2(2mZe^2a_0z/\hbar^2)^{1/2},$$

$$(3.38.10b)$$

and hence (3.38.7) gives

$$\rho \approx mZe^2a_0z/(-\eta\hbar^2). \qquad (3.38.11)$$

According to sections 14.1.3–14.1.6 and 14.1.8 in Abramowitz and Stegun (1965) and (3.38.11) we have, for sufficiently small values of ρ and $2\pi\eta \ll -1$,

$$F_0(\eta, \rho) \approx C_0(\eta)\rho = \left(\frac{2\pi\eta}{e^{2\pi\eta} - 1}\right)^{1/2} \rho \approx (-2\pi\eta)^{1/2} \rho$$

$$\approx (-2\pi\eta)^{1/2} \frac{mZe^2a_0z}{-\eta\hbar^2} = \frac{mZe^2a_0}{\hbar^2} \left(-\frac{2\pi}{\eta}\right)^{1/2} z. \quad (3.38.12)$$

Since, furthermore, according to (3.38.11) $(d\rho/dz)^{-1/2} \approx [-\eta \hbar^2/(mZe^2a_0)]^{1/2}$, we obtain from (3.38.5) and (3.38.12), for small values of ρ and $2\pi\eta \ll -1$,

$$\psi \approx A[-\eta\hbar^2/(mZe^2a_0)]^{1/2}(mZe^2a_0/\hbar^2)(-2\pi/\eta)^{1/2}z = A(2\pi mZe^2a_0/\hbar^2)^{1/2}z.$$
$$(3.38.13)$$

Putting

$$A = [\hbar^2/(2\pi mZe^2a_0)]^{1/2}, \qquad (3.38.14)$$

we fulfil the requirement that $\psi(z)/z$ tends to unity as z tends to zero.

It now remains to obtain an expression for $\psi(z)$ when z lies far away from the origin, but still in the classically allowed region that contains the origin. For $F_0(\eta, \rho)$ we have the asymptotic formula

$$F_0(\eta, \rho) \sim \cos[\rho - \eta \ln(2\rho) + \arg \Gamma(1 + i\eta) - \pi/2], \quad \rho \to +\infty. \quad (3.38.15)$$

Since

$$\int_0^\rho \left(1 - \frac{2\eta}{\rho}\right)^{1/2} d\rho = \rho\left(1 - \frac{2\eta}{\rho}\right)^{1/2} + \eta \ln \frac{(1 - 2\eta/\rho)^{1/2} - 1}{(1 - 2\eta/\rho)^{1/2} + 1}$$

$$= \rho - \eta \ln(2\rho) - \eta + \eta \ln(-\eta) + O(1/\rho), \quad \rho \to +\infty,$$
$$(3.38.16)$$

i.e.,

$$\lim_{\rho\to+\infty} \left[\rho - \eta \ln(2\rho) - \int_0^\rho (1 - 2\eta/\rho)^{1/2} d\rho\right] = \eta - \eta \ln(-\eta), \quad (3.38.16')$$

we can write the asymptotic formula (3.38.15) for $F_0(\eta, \rho)$ as

$$F_0(\eta, \rho) \sim \cos\left[\int_0^\rho (1 - 2\eta/\rho)^{1/2} d\rho + \Delta\right], \quad \rho \to +\infty, \quad (3.38.17)$$

where

$$\Delta = \arg \Gamma(1 + i\eta) + \eta - \eta \ln(-\eta) - \pi/2. \qquad (3.38.18)$$

In order to rewrite (3.38.17) along with (3.38.18) as a first-order phase-integral formula that is also valid for finite values of ρ, we note that, since $F_0(\eta, \rho)$ is a solution of (3.38.6), the first-order phase-integral formula for $F_0(\eta, \rho)$ that corresponds to the asymptotic formula (3.38.17) must be of the form

$$F_0(\eta, \rho) \approx \text{const} \times (1 - 2\eta/\rho)^{-1/4} \cos \left[\int_0^\rho (1 - 2\eta/\rho)^{1/2} d\rho + \text{const} \right].$$

$$(3.38.19)$$

Letting $\rho \to +\infty$, we can identify (3.38.19) with (3.38.17) and determine the unknown constants in (3.38.19). Thus we get the approximate formula

$$F_0(\eta, \rho) \approx (1 - 2\eta/\rho)^{-1/4} \cos \left[\int_0^\rho (1 - 2\eta/\rho)^{1/2} d\rho + \Delta \right], \quad (3.38.20)$$

which is valid unless $\rho \, (> 0)$ lies too close to the origin.

From (3.38.5), (3.38.14), (3.38.20), (3.38.7) and (3.38.18) we now obtain for the wave function $\psi(z)$ that is normalized such that $\psi(z)/z \to 1$ as $z \to 0$ the phase-integral formula

$$\psi \approx \left[\hbar^2/(2\pi \, m Z e^2 a_0) \right]^{1/2} Q^{-1/2}(z)$$

$$\times \cos \left[\int_0^z Q(z) dz + \arg \Gamma(1 + i\eta) + \eta - \eta \ln(-\eta) - \pi/2 \right]. \quad (3.38.21)$$

According to eq. (5) on p. 32 in Luke (1969) we have the asymptotic formula

$$\ln \Gamma(1 \pm i\eta) \sim (1/2 \pm i\eta) \ln(1/2 \pm i\eta) - (1/2 \pm i\eta) + \ln(2\pi)^{1/2},$$

$$|1/2 \pm i\eta| \gg 1, \quad -\pi + \varepsilon < \arg(1/2 \pm i\eta) < \pi - \varepsilon, \quad (3.38.22)$$

where ε is a fixed, arbitrarily small, positive number. By means of (3.38.22) we obtain

$$\frac{1}{2i} \ln \frac{\Gamma(1 + i\eta)}{\Gamma(1 - i\eta)} \sim \frac{1}{2i} [(1/2 + i\eta) \ln(1/2 + i\eta) - (1/2 - i\eta) \ln(1/2 - i\eta)] - \eta.$$

$$(3.38.23)$$

When η is real, this formula can be written as

$$\arg \Gamma(1 + i\eta) \sim \eta \ln(\eta^2 + 1/4)^{1/2} - \eta + \tfrac{1}{2}\arctan(2\eta) \approx \eta \ln(-\eta) - \eta - \pi/4,$$

$$(3.38.24)$$

where to obtain the last expression we have used the assumption that η is negative and sufficiently large. Inserting (3.38.24) into (3.38.21), we obtain the approximate formula

$$\psi = [\hbar^2/(2\pi m Z e^2 a_0)]^{1/2} Q^{-1/2}(z) \cos\left[\int_0^z Q(z)dz - 3\pi/4\right]. \quad (3.38.25)$$

One obtains the generalization of this formula to an arbitrary-order approximation by replacing $Q(z)$ by $q(z)$ and generalizing the integral in (3.38.25) according to (2.2.14) and (2.2.15).

A first-order formula that is more general than (3.38.25) was derived by Fröman and Fröman (1965): their eq. (7.28). In this formula, which is valid when $2l + 1$ is a small integer, where l corresponds to the angular momentum quantum number, the base function is chosen to be $Q(z) = [R(z) + l(l + 1)/z^2]^{1/2}$; cf. (3.38.8). The formula was derived without using the comparison equation technique. Also this formula can be generalized to an arbitrary order of the phase-integral approximation generated from an unspecified base function.

3.39 Quantization condition, and value of the normalized wave function at the origin expressed in terms of the level density, for an *s*-electron in a single-well potential with a strong attractive Coulomb singularity at the origin

Use formula (3.38.25) to derive the quantization condition and the value at the origin of the normalized wave function for an *s*-electron (i.e., an electron with the angular momentum quantum number l equal to zero) in the radial single-well potential of an atom or ion with large nuclear charge number Z.

Solution. The normalized wave function for an *s*-electron is (cf. (3.38.3a,b))

$$\Psi(r) = \frac{u(r)}{(4\pi)^{1/2}r} = \frac{\psi(z)/z}{\left(4\pi a_0^3\right)^{1/2}}, \quad (3.39.1)$$

where $s = 0, 1, 2, \ldots$ is the quantum number of the radial *s*-electron wave function $\psi(z)$, which we can assume to be real without any essential restriction. This wave function satisfies the radial Schrödinger equation (cf. (3.38.3a) and (3.38.4a,b))

$$d^2\psi/dz^2 + R(z)\psi = 0, \quad (3.39.2a)$$

$$R(z) = 2ma_0^2[E - V(r)]/\hbar^2, \quad (3.39.2b)$$

$$z = r/a_0, \quad (3.39.2c)$$

and according to a remark below (3.38.3a,b) it is normalized such that

$$\int_0^\infty \psi^2(z)dz = 1.$$
(3.39.3)

If we choose

$$Q(z) = R^{1/2}(z),$$
(3.39.4)

which is convenient in this problem, the phase-integral approximation breaks down for small values of z. However, when the nuclear charge number Z is sufficiently large, and z lies sufficiently far away from, but still in the same classically allowed region as the origin, the radial wave function $\psi_1(z)$ that is equal to zero at the origin for any value of the energy can according to (3.38.25) be represented by the first-order phase-integral formula

$$\psi_1(z) = N_1 Q^{-1/2}(z) \cos \left[\int_0^z Q(z)dz - 3\pi/4 \right].$$
(3.39.5)

Here $Q(z)$ is chosen to be positive in the classically allowed region, and N_1 is a constant normalization factor. The physically acceptable wave function $\psi(z)$ that in the classically allowed region is given by (3.38.25) behaves such that $\psi(z)/z \to 1$ as z tends to zero, and therefore the physically acceptable wave function that in the classically allowed region is given by (3.39.5) behaves as z tends to zero such that

$$\lim_{z \to 0} \psi_1(z)/z = N_1(2\pi \, mZe^2 a_0/\hbar^2)^{1/2}.$$
(3.39.6)

According to the connection formula (3.10.10′) the particular solution $\psi_2(z)$ that tends to zero as $z \to +\infty$ for any value of the energy is, in the classically allowed region of the radial single-well potential under consideration, given by the phase-integral formula

$$\psi_2(z) = N_2 Q^{-1/2}(z) \cos \left[\int_z^t Q(z)dz - \pi/4 \right]$$

$$= N_2 Q^{-1/2}(z) \cos \left[\int_0^z Q(z)dz - \int_0^t Q(z)dz + \pi/4 \right],$$
(3.39.7)

where t is the turning point, and N_2 is a constant normalization factor.

Requiring that $\psi_1(z) = \psi_2(z)$, we obtain from (3.39.5) and (3.39.7)

$$\int_0^t Q(z)dz = (s+1)\pi, \qquad (3.39.8a)$$

$$N_2 = (-1)^s N_1, \qquad (3.39.8b)$$

where s is an integer, which must be non-negative. This is the quantization condition that determines the energy levels E_s.

The normalization integral for the bound-state radial wave function $\psi(z, E_s)$ that coincides with $\psi_1(z)$ and $\psi_2(z)$ when $E = E_s$ is according to (3.36.10) and (3.38.3a,b)

$$\int_0^\infty \psi^2(z, E_s)dz = \frac{\hbar^2}{2ma_0^2} \left\{ \frac{\partial}{\partial E} [\psi_1(z)\psi_2'(z) - \psi_2(z)\psi_1'(z)] \right\}_{E=E_s}, \qquad (3.39.9)$$

where the prime indicates differentiation with respect to z. Using (3.39.5), (3.39.7) and (3.39.8b) we easily find that

$$\psi_1(z)\psi_2'(z) - \psi_2(z)\psi_1'(z) = (-1)^{s+1} N_1^2 \sin\left[\int_0^t Q(z)dz \right], \qquad (3.39.10)$$

and hence, with the aid of (3.39.8a),

$$\left\{ \frac{\partial}{\partial E} [\psi_1(z)\psi_2'(z) - \psi_2(z)\psi_1'(z)] \right\}_{E=E_s}$$

$$= (-1)^{s+1} \left\{ N_1^2 \cos\left[\int_0^t Q(z)dz \right] \frac{\partial}{\partial E} \int_0^t Q(z)dz \right\}_{E=E_s}$$

$$= \left(N_1^2 \frac{\partial}{\partial E} \int_0^t Q(z)dz \right)_{E=E_s}. \qquad (3.39.11)$$

From (3.39.9) and (3.39.11) we obtain

$$\int_0^\infty \psi^2(z, E_s)dz = \frac{\hbar^2}{2m\,a_0^2} \left[N_1^2 \frac{\partial}{\partial E} \int_0^t Q(z)dz \right]_{E=E_s}, \qquad (3.39.12)$$

which with due regard to the normalization condition (3.39.3) yields

$$N_1 = \left[\frac{\hbar^2}{2ma_0^2} \frac{\partial}{\partial E} \int_0^t Q(z)dz \right]_{E=E_s}^{-1/2}, \qquad (3.39.13)$$

if we assume N_1 to be positive. Since $\psi(z) \equiv \psi_1(z)$ for $E = E_s$, we obtain from (3.39.1), (3.39.6) and (3.39.13)

$$\Psi(0) = \lim_{z \to 0} \frac{\psi_1(z)/z}{(4\pi a_0^3)^{1/2}} = \left(\frac{mZe^2}{2\hbar^2 a_0^2}\right)^{1/2} \quad N_1 = \frac{meZ^{1/2}}{\hbar^2} \left[\frac{\partial}{\partial E} \int_0^t Q(z)dz\right]_{E=E_s}^{-1/2}.$$

(3.39.14)

From the quantization condition (3.39.8a) we obtain the approximate formula

$$\left[\frac{\partial}{\partial E} \int_0^t Q(z)dz\right]_{E=E_s} = \pi \left(\frac{ds}{dE}\right)_{E=E_s},$$

(3.39.15)

and therefore (3.39.14) can be expressed in terms of the level density ds/dE as

$$\Psi(0) = \frac{me}{\hbar^2} \left[\frac{Z}{\pi(ds/dE)_{E=E_s}}\right]^{1/2}.$$

(3.39.16)

3.40 Expectation value of an unspecified function $f(z)$ for a non-relativistic particle in a bound state

Derive an exact formula, not involving the wave function, for the expectation value of an unspecified function $f(z)$ for a non-relativistic quantal particle in a bound state of a *general* potential $V(z)$. Use this formula to obtain a phase-integral formula for the expectation value of $f(z)$ in a smooth *single-well* potential.

Solution. Consider the auxiliary differential equation

$$d^2\psi/dz^2 + R(z, E, \kappa)\psi = 0,$$ (3.40.1a)
$$R(z, E, \kappa) = 2m[E - V(z) - \kappa f(z)]/\hbar^2,$$ (3.40.1b)

where $V(z)$ is the effective potential, and κ is a small parameter. Putting $\kappa = 0$ in (3.40.1a,b), one obtains the original differential equation. We assume that the auxiliary potential $V(z) + \kappa f(z)$ appearing here has bound states for all sufficiently small values of κ. The eigenfunction ψ_s (which can be assumed to be real) and the eigenvalue E_s of the auxiliary differential equation, obtained from the requirement that ψ_s is equal to zero for $z = a_1$ and for $z = a_2$, depend on κ. Putting $\psi = \psi_s$ and $E = E_s(\kappa)$ and differentiating the auxiliary differential equation (3.40.1a) partially with respect to κ, we obtain

$$\frac{\partial^2}{\partial z^2}\frac{\partial \psi_s}{\partial \kappa} + R\frac{\partial \psi_s}{\partial \kappa} + \frac{\partial R}{\partial \kappa}\psi_s = 0.$$ (3.40.2)

Multiplying this equation by ψ_s and the auxiliary Schrödinger equation (3.40.1a), with ψ replaced by ψ_s, by $\partial\psi_s/\partial\kappa$, we obtain

$$\psi_s\frac{\partial^2}{\partial z^2}\frac{\partial\psi_s}{\partial\kappa} + R\psi_s\frac{\partial\psi_s}{\partial\kappa} + \frac{\partial R}{\partial\kappa}\psi_s^2 = 0, \tag{3.40.3a}$$

$$\frac{\partial\psi_s}{\partial\kappa}\frac{\partial^2\psi_s}{\partial z^2} + R\psi_s\frac{\partial\psi_s}{\partial\kappa} = 0. \tag{3.40.3b}$$

Subtraction of these two equations gives

$$\frac{\partial}{\partial z}\left(\psi_s\frac{\partial}{\partial z}\frac{\partial\psi_s}{\partial\kappa} - \frac{\partial\psi_s}{\partial\kappa}\frac{\partial\psi_s}{\partial z}\right) + \frac{\partial R}{\partial\kappa}\psi_s^2 = 0. \tag{3.40.4}$$

Since ψ_s is equal to zero for $z = a_1$ and for $z = a_2$, the partial derivative $\partial\psi_s/\partial\kappa$ is also equal to zero for $z = a_1$ and $z = a_2$. Therefore from (3.40.4) we obtain

$$\int_{a_1}^{a_2}\frac{\partial R}{\partial\kappa}\psi_s^2\,dz = 0, \tag{3.40.5}$$

i.e., with the aid of (3.40.1b) with $E = E_s(\kappa)$,

$$\int_{a_1}^{a_2}\left[\frac{dE_s(\kappa)}{d\kappa} - f(z)\right]\psi_s^2\,dz = 0, \tag{3.40.6}$$

i.e.,

$$\frac{\displaystyle\int_{a_1}^{a_2} f(z)\psi_s^2\,dz}{\displaystyle\int_{a_1}^{a_2}\psi_s^2\,dz} = \frac{dE_s(\kappa)}{d\kappa}. \tag{3.40.7}$$

Here ψ_s is an eigenfunction of the auxiliary differential equation (3.40.1a), but for $\kappa = 0$ it is also an eigenfunction of the original differential equation. Therefore we obtain for the expectation value $\langle f(z)\rangle$ of $f(z)$ the exact formula

$$\langle f(z)\rangle = [dE_s(\kappa)/d\kappa]_{\kappa=0}, \tag{3.40.8}$$

which is the Hellmann–Feynman formula for the auxiliary differential equation (3.40.1a,b), here utilized when $\kappa = 0$. This formula is valid even if $V(z)$ is not a single-well potential.

From the exact formula (3.40.8) one can obtain an approximate, but often very accurate, phase-integral formula by using the quantization condition for the auxiliary differential equation (3.40.1a,b). If $V(z) + \kappa f(z)$ is a smooth *single-well* potential, the quantization condition is

$$\frac{1}{2}\int_\Lambda q(z, E_s, \kappa)dz = (s + 1/2)\pi, \qquad s = 0, 1, 2, \ldots, \qquad (3.40.9)$$

where Λ is a closed contour of integration that encircles the turning points t' and t'', i.e., the relevant zeros of $Q^2(z, E_s, \kappa)$ on the real z-axis, in the direction for which the integral is positive. Differentiating (3.40.9) partially with respect to κ, for fixed quantum number s, we obtain

$$\frac{\partial}{\partial\kappa}\int_\Lambda q(z, E_s, \kappa)dz = 0. \qquad (3.40.10)$$

Under the assumption that $R(z, E, \kappa) - Q^2(z, E, \kappa)$ is independent of κ, N. Fröman (1974) and P. O. Fröman (1974) proved the approximate, but in general very accurate, formula

$$\frac{\partial}{\partial\kappa}\int_\Lambda q(z, E, \kappa)dz = \int_\Lambda \frac{1}{2}\frac{\partial R(z, E, \kappa)}{\partial\kappa}\frac{dz}{q(z, E, \kappa)}, \qquad (3.40.11)$$

where on the right-hand side Λ must for higher-order approximations encircle not only the turning points t' and t'', but also the zeros of $q(z, E_s)$ located in the neighbourhood of t' and t'' and mentioned below (2.2.11). A particular case of this formula is (3.37.11).

Assuming thus that $R(z, E, \kappa) - Q^2(z, E, \kappa)$ is independent of κ, we obtain from (3.40.10), (3.40.11) and (3.40.1b)

$$\int_\Lambda \left[\frac{dE_s(\kappa)}{d\kappa} - f(z)\right]\frac{dz}{q(z, E_s, \kappa)} = 0,$$

i.e.,

$$\frac{dE_s(\kappa)}{d\kappa} = \frac{\displaystyle\int_\Lambda \frac{f(z)dz}{q(z, E_s, \kappa)}}{\displaystyle\int_\Lambda \frac{dz}{q(z, E_s, \kappa)}}. \qquad (3.40.12)$$

Inserting (3.40.12) into (3.40.8) and putting $q(z, E_s, 0) = q(z)$, we obtain (cf. (2.2.14) and (2.2.15))

$$\langle f(z)\rangle = \frac{\displaystyle\int_\Lambda \frac{f(z)dz}{q(z)}}{\displaystyle\int_\Lambda \frac{dz}{q(z)}} = \frac{\displaystyle\int_{(t')}^{(t'')} \frac{f(z)dz}{q(z)}}{\displaystyle\int_{(t')}^{(t'')} \frac{dz}{q(z)}}. \qquad (3.40.13)$$

Since $[E_s - V(z)]^{1/2}$ is proportional to the classical velocity of the particle, the first order of (3.40.13), with the base function $Q(z, E_s)$ chosen to be equal to $R^{1/2}(z, E_s)$, is obviously the same as the formula for the classical time average value of $f(z)$; see Section 3.26.

By assuming κ to be sufficiently small, we can make $\kappa f(z)$ arbitrarily small over as large a part of the real axis as desired. Hence, using simple physical arguments, we realize that the formulas derived in the present problem can be used even if the auxiliary potential $V(z) + \kappa f(z)$ for $\kappa \neq 0$ does not strictly fulfil the condition of being a potential with truly bound states.

In the literature the first order of (3.40.13) is usually derived by inserting the first-order WKB approximation for the wave function into the integral defining the expection value of $f(z)$, then simplifying the resulting expression by neglecting contributions from the classically forbidden regions, and replacing the square of the cosine in the classically allowed region by its average value 1/2. In view of those approximations one would expect a rather rough formula, and the high accuracy of the formula has puzzled many authors. So also has the fact that attempts to improve the formula by using a uniform approximation or Airy functions for the wave function through the turning points failed. The derivation given in this problem shows that one may, quite generally, expect the usefulness of the phase-integral formula (3.40.13) for expectation values to be about the same as that for the phase-integral quantization condition, which is known to be very accurate for important classes of physical problems. Our derivation thus explains why the first-order formula is remarkably accurate and why the above-mentioned attempts to improve it failed.

For the background of this problem and for applications of the resulting formula we refer the reader to N. Fröman (1974, 1978a, 1980), P. O. Fröman (1974), Fröman, Fröman and Karlsson (1979), Thidé (1980), Fröman and Fröman (1985b), Paulsson and N. Fröman (1985), Paulsson (1985), P. O. Fröman, Hökback, Walles and Yngve (1985), Linnæus and Düring (1985) and Yngve and Linnæus (1986).

3.41 Some cases in which the phase-integral expectation value formula yields the expectation value exactly in the first-order approximation

Use the fact that, with a convenient choice of the base function, the phase-integral quantization condition is exact for the harmonic oscillator and the hydrogen atom to show, without explicit calculation, that the first order of the formula (3.40.13) yields the following expection values exactly:

(a) $\langle z^0 \rangle$, $\langle z^1 \rangle$ and $\langle z^2 \rangle$ for the harmonic oscillator;
(b) $\langle z^0 \rangle$, $\langle z^{-1} \rangle$ and $\langle z^{-2} \rangle$ for the hydrogen atom.

Solution. The first order of the quantization condition (3.40.9), i.e.,

$$\int_{t'}^{t''} Q(z, E, \kappa)\,dz = (s + 1/2)\pi, \qquad (3.41.1)$$

is exact, when in case (a)

$$Q^2(z, E, \kappa) = R(z, E, \kappa) = (2m/\hbar^2)\left[E - \frac{1}{2}m\omega^2 z^2 - \kappa z^n\right], \qquad n = 0, 1, 2, \qquad (3.41.2)$$

and when in case (b)

$$\begin{aligned} Q^2(z, E, \kappa) &= R(z, E, \kappa) - 1/(4z^2) \\ &= (2m/\hbar^2)[E + e^2/z - \hbar^2(l + 1/2)^2/(2mz^2) - \kappa z^{-n}], \end{aligned}$$

$$n = 0, 1, 2. \quad (3.41.3)$$

One can easily verify that the formula (3.40.11), and hence also (3.40.12), is exact for the first-order approximation when $R(z, E, \kappa) - Q^2(z, E, \kappa)$ is independent of κ. Therefore (3.40.13) yields the expectation values exactly in both case (a) and case (b).

3.42 Expectation value of the kinetic energy of a non-relativistic particle in a bound state. Verification of the virial theorem

Determine the expectation value of the kinetic energy for a bound state of a non-relativistic particle in a single-well potential. Apply the resulting formula to calculate the expectation value of the kinetic energy of a non-relativistic particle in the harmonic oscillator potential $V(z) = m\omega^2 z^2/2$ and verify the virial theorem for that system.

Solution. The Schrödinger equation is

$$d^2\psi/dz^2 + R(z)\psi = 0, \qquad (3.42.1a)$$
$$R(z) = 2m[E - V(z)]/\hbar^2. \qquad (3.42.1b)$$

The kinetic energy is

$$T(z) = E - V(z) = \hbar^2 R(z)/(2m), \qquad (3.42.2)$$

and according to (3.40.13) its quantal expectation value is

$$\langle T \rangle = \frac{\displaystyle\int_\Lambda \frac{T(z)dz}{q(z)}}{\displaystyle\int_\Lambda \frac{dz}{q(z)}} = \frac{\hbar^2}{2m} \frac{\displaystyle\int_\Lambda \frac{R(z)dz}{q(z)}}{\displaystyle\int_\Lambda \frac{dz}{q(z)}}. \tag{3.42.3}$$

Choosing the square of the base function to be

$$Q^2(z) = R(z), \tag{3.42.4}$$

one obtains from (3.42.3), when the first-order approximation is used,

$$\langle T \rangle = \frac{\hbar^2}{2m} \frac{\displaystyle\int_{(t')}^{(t'')} Q(z)dz}{\displaystyle\int_{(t')}^{(t'')} \frac{dz}{Q(z)}}. \tag{3.42.5}$$

In the first-order approximation the quantization condition (3.40.9) with $\kappa = 0$ is

$$\int_{t'}^{t''} Q(z, E_s)dz = (s + 1/2)\pi, \tag{3.42.6}$$

where $Q(z)$ is positive in the classically allowed region. Differentiating (3.42.6) with respect to s, which is considered to be a continuous parameter, and using (3.42.4) and (3.42.1b), we get

$$\int_{t'}^{t''} \frac{dz}{Q(z, E_s)} = \frac{\pi \hbar^2}{m \, dE_s/ds}. \tag{3.42.7}$$

Inserting (3.42.6) and (3.42.7) into (3.42.5), we obtain

$$\langle T \rangle = \frac{1}{2}\left(s + \frac{1}{2}\right)\frac{dE_s}{ds}. \tag{3.42.8}$$

The energy eigenvalues of a particle in the harmonic oscillator potential $V(z) = m\omega^2 z^2/2$ are $E_s = (s + 1/2)\hbar\omega$, $s = 0, 1, 2, \ldots$. From (3.42.8) we therefore obtain

$$\langle T \rangle = \frac{1}{2}\left(s + \frac{1}{2}\right)\hbar\omega = \frac{1}{2}E_s, \tag{3.42.9}$$

in agreement with the exact value of the quantity in question and with the virial theorem.

3.43 Phase-integral calculation of quantal matrix elements

Use (3.40.13) to derive a phase-integral formula for the quantal matrix element of $f(z)$ between the ground state and an excited state in a single-well potential.

Solution. Consider the Schrödinger equation

$$d^2\psi/dz^2 + R(z, E)\psi = 0, \tag{3.43.1a}$$

$$R(z, E) = (2m/\hbar^2)[E - V(z)]. \tag{3.43.1b}$$

The function $V(z)$ may be the actual physical potential or an effective potential. Thus, if we are concerned with the radial Schrödinger equation, $V(z)$ also includes the centrifugal term.

Let $\psi(z, E_s)$ and $\psi(z, E_{s'})$ be two normalized eigenfunctions (real on the real z-axis) corresponding to the eigenvalues E_s and $E_{s'}$. If $\psi(z, E_s)$ is the wave function of the ground state ($s = 0$), this wave function is different from zero on the real z-axis. Therefore we can write the matrix element of $f(z)$ as an expectation value:

$$\langle s|f(z)|s'\rangle = \int \psi(z, E_s) f(z)\psi(z, E_{s'})dz$$

$$= \int \psi(z, E_s)\frac{f(z)\psi(z, E_{s'})}{\psi(z, E_s)}\psi(z, E_s)dz = \left\langle s \left| \frac{f(z)\psi(z, E_{s'})}{\psi(z, E_s)} \right| s \right\rangle, \tag{3.43.2}$$

where the integration is to be performed along the part of the real z-axis that is appropriate to the range of the physical variable z. Using (3.40.13), we obtain from (3.43.2) the approximate formula

$$\langle s|f(z)|s'\rangle = \frac{\displaystyle\int_\Lambda \frac{f(z)\,\psi(z, E_{s'})}{\psi(z, E_s)}\frac{dz}{q(z, E_s)}}{\displaystyle\int_\Lambda \frac{dz}{q(z, E_s)}}. \tag{3.43.3}$$

When $Q(z, E)$ is chosen to be positive on the upper lip of the cut between the turning points t' and t'' ($> t'$), and z lies sufficiently far away from the classically allowed region, the normalized eigenfunction $\psi(z, E_s)$, which is positive on the real z-axis to the left of t', is given by the approximate formula (3.37.14a,b), i.e.,

$$\psi(z, E_s) = \frac{\exp(i\pi/4)q^{-1/2}(z, E_s)\exp[-iw(z, E_s)]}{\left[\displaystyle\int_\Lambda dz/q(z, E_s)\right]^{1/2}}, \qquad (3.43.4a)$$

$$w(z, E_s) = \int_{(t')}^{z} q(z, E_s)dz. \qquad (3.43.4b)$$

The contour of integration Λ in (3.43.4a) encircles the turning points and the associated zeros of $q(z, E_s)$ in the negative direction; see the comment below (2.2.11). From (3.43.4a) and the corresponding formula with E_s replaced by $E_{s'}$ we obtain

$$\frac{\psi(z, E_{s'})}{\psi(z, E_s)} = \frac{\left[\displaystyle\int_\Lambda dz/q(z, E_s)\right]^{1/2} q^{1/2}(z, E_s)}{\left[\displaystyle\int_\Lambda dz/q(z, E_{s'})\right]^{1/2} q^{1/2}(z, E_{s'})} \exp\{i[w(z, E_s) - w(z, E_{s'})]\}.$$

$$(3.43.5)$$

Inserting (3.43.5) into (3.43.3), we obtain the approximate formula

$$\langle s|f(z)|s'\rangle = \frac{\displaystyle\int_\Lambda \frac{\exp\{i[w(z, E_s) - w(z, E_{s'})]\}}{q^{1/2}(z, E_s)q^{1/2}(z, E_{s'})} f(z)dz}{\left[\displaystyle\int_\Lambda \frac{dz}{q(z, E_s)}\right]^{1/2}\left[\displaystyle\int_\Lambda \frac{dz}{q(z, E_{s'})}\right]^{1/2}}. \qquad (3.43.6)$$

For the sake of simplicity we have derived this formula under the assumption that s is the quantum number of the ground state, i.e., that $s = 0$. The formula is also valid when s is not the quantum number of the ground state, but the proof is then tricky; see P. O. Fröman (2000). The formula is, in fact, expected to be more accurate when $E_s > E_{s'}$ than when $E_s < E_{s'}$, and its accuracy increases as $E_s - E_{s'} (> 0)$ decreases. For further details we refer the reader to Fröman and Fröman (1977) and P. O. Fröman (2000). Gustafson and Lindahl (1977) have presented an efficient method for numerical calculation of the integral in the numerator of (3.43.6).

A potential in which the states are unbound can be considered as the limiting case of a potential in which the states are bound. In that way one can obtain from (3.43.6) a formula for matrix elements associated with unbound states; see Fröman, Fröman and Karlsson (1979). In some applications it has turned out that this formula may sometimes be superior to numerical methods in efficiency and accuracy. For certain matrix elements one had not been able to obtain more than four or five correct digits with numerical methods, but Fröman, Fröman and Karlsson (1979) obtained between 12 and 20 correct digits for these matrix elements when their phase-integral formula was used in higher-order approximations. With the aid of exact relations between different matrix elements they were able to check that all these digits are correct.

For the background of this problem and for applications of the resulting formulas we refer the reader to Fröman and Fröman (1977), Fröman, Fröman and Karlsson (1979), N. Fröman (1980), Paulsson and N. Fröman (1985), Paulsson (1985), P. O. Fröman, Hökback, Walles and Yngve (1985), Yngve and Linnæus (1986), Karlsson and Jedrzejek (1987), Streszewski and Jedrzejek (1988) and P. O. Fröman (2000).

3.44 Connection formula for a complex potential barrier

Derive the connection formula associated with a complex potential barrier.

Solution. By a *complex potential barrier* we mean a cluster of two transition zeros with the property that these two transition zeros are *not joined by an anti-Stokes line*. Whether a cluster of two transition zeros is to be characterized as a complex potential barrier or not may depend on the order of approximation used since the anti-Stokes lines in general change slightly when the order changes.

The definition of what we mean by saying that an arbitrary-order Stokes or anti-Stokes line emerges from a transition point of odd order has been explained at the end of Section 2.3.5. Disregarding cases in which anti-Stokes lines emerge from the transition zeros delimiting the complex barrier and end at transition points that are not associated with the barrier, we realize that the three anti-Stokes lines, emerging from each one of the two transition zeros of the complex potential barrier, all proceed towards infinity; see Fig. 3.44.1.

The wave function is single-valued when z encircles one or more transition zeros. This fact was used in Section 3.7 to calculate Stokes constants and in Section 3.28 to derive the quantization condition for a particle in a single-well potential. In the solution of the present problem the same fact will be used to obtain information on the Stokes constants pertaining to the four sectors denoted, like their associated Stokes constants, by S', S'', $\bar{S}'$ and $\bar{S}''$ in Fig. 3.44.1. These sectors are delimited by anti-Stokes lines emerging from the transition zeros of the complex barrier.

For any z we write

$$\psi(z) = a_1(z)q^{-1/2}(z)\exp\left[+i\int_{(t')}^{z} q(z)dz\right] + a_2(z)q^{-1/2}(z)\exp\left[-i\int_{(t')}^{z} q(z)dz\right].$$

$$(3.44.1)$$

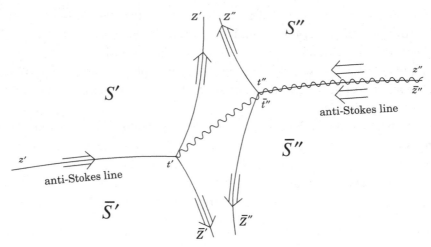

Figure 3.44.1 The double arrows indicate the directions in which $\operatorname{Re}\int^{z} q\,(z)\,dz$ increases along the anti-Stokes lines emerging from the transition zeros. There is a cut (wavy line) joining t' and t'' and proceeding from t'' to infinity along an anti-Stokes line emerging from t''. When the potential barrier is real, and t' is the left-hand transition zero, the phase of $Q(z)$ is the same as in Fröman and Fröman (1970).

When z' and Z' lie sufficiently far away from t' and t'', the a-coefficients at z' and Z' are related according to the approximate formula (see Fig. 3.44.1 and (2.3.20a))

$$\begin{pmatrix} a_1(Z') \\ a_2(Z') \end{pmatrix} = \begin{pmatrix} 1 & S' \\ 0 & 1 \end{pmatrix} \begin{pmatrix} a_1(z') \\ a_2(z') \end{pmatrix}, \tag{3.44.2}$$

where S' is a Stokes constant, which according to Fig. 3.44.1, (3.7.7b), (3.7.8) and (2.3.12c) is approximately equal to $-i$, when t' and t'' are well separated. Since the imaginary part of $\int_{Z'}^{Z''} q(z)dz$ is equal to the imaginary part of $\int_{(t')}^{(t'')} q(z)dz$, we realize that

$$\left| \exp\left[\pm i \int_{Z'}^{Z''} q(z)dz \right] \right| = \left| \exp\left[\pm i \int_{(t')}^{(t'')} q(z)dz \right] \right|.$$

Hence, this absolute value remains constant and thus finite, while the μ-integral $\left|\int_{Z'}^{Z''} |\varepsilon(z)q(z)dz|\right|$ diminishes to a very small value, as Z' and Z'' both move away from t' and t'' on the anti-Stokes lines in question; see Fig. 3.44.1. According to (2.3.7) and the basic estimates (2.3.17a–d) we therefore have approximately

$$\begin{pmatrix} a_1(Z'') \\ a_2(Z'') \end{pmatrix} = \begin{pmatrix} a_1(Z') \\ a_2(Z') \end{pmatrix}. \tag{3.44.3}$$

The approximate formula that relates the a-coefficients at Z'' and z'' is (see Fig. 3.44.1 and (2.3.20b))

$$\begin{pmatrix} a_1(z'') \\ a_2(z'') \end{pmatrix} = \begin{pmatrix} 1 & 0 \\ S'' & 1 \end{pmatrix} \begin{pmatrix} a_1(Z'') \\ a_2(Z'') \end{pmatrix}, \tag{3.44.4}$$

where S'' is a Stokes constant. Recalling Fig. 3.44.1, (3.7.3b), (3.7.6) and (2.3.12c) and noting that the lower limit of integration in the integral in (3.44.1) is t', we realize by means of (2.3.28) together with (2.3.27b) that, when t' and t'' are well separated, S'' is approximately equal to $-i \exp(2K)$, where

$$K = i \int\limits_{(t')}^{(t'')} q(z)dz = \frac{1}{2i} \int\limits_{\Lambda} q(z)dz, \tag{3.44.5}$$

Λ being a closed contour encircling t' and t'' along which the integration is performed in the positive direction; cf. (2.5.12a,b). From (3.44.4), (3.44.3) and (3.44.2) we obtain

$$\begin{pmatrix} a_1(z'') \\ a_2(z'') \end{pmatrix} = \begin{pmatrix} a_1(z') + S'a_2(z') \\ S''a_1(z') + (1 + S'S'')a_2(z') \end{pmatrix}, \tag{3.44.6}$$

and from (3.44.1) and (3.44.6) we obtain

$$\psi(z'') = [a_1(z') + S'a_2(z')]q^{-1/2}(z'') \exp\left[+i \int\limits_{(t')}^{z''} q(z)dz \right]$$

$$+ [S''a_1(z') + (1 + S'S'')a_2(z')]q^{-1/2}(z'') \exp\left[-i \int\limits_{(t')}^{z''} q(z)dz \right]. \tag{3.44.7}$$

Recalling (2.3.20b), we realize that

$$\begin{pmatrix} a_1(\bar{Z}') \\ a_2(\bar{Z}') \end{pmatrix} = \begin{pmatrix} 1 & 0 \\ \bar{S}' & 1 \end{pmatrix} \begin{pmatrix} a_1(z') \\ a_2(z') \end{pmatrix}, \tag{3.44.8}$$

where $\bar{S}'$ is a Stokes constant, which, when t' and t'' are well separated, is approximately equal to i according to Fig. 3.44.1, (3.7.3b) and (3.7.6). In analogy to (3.44.3) we have the approximate formula

$$\begin{pmatrix} a_1(\bar{Z}'') \\ a_2(\bar{Z}'') \end{pmatrix} = \begin{pmatrix} a_1(\bar{Z}') \\ a_2(\bar{Z}') \end{pmatrix}. \tag{3.44.9}$$

According to (2.3.20a) we have

$$\begin{pmatrix} a_1(\bar{z}'') \\ a_2(\bar{z}'') \end{pmatrix} = \begin{pmatrix} 1 & \bar{S}'' \\ 0 & 1 \end{pmatrix} \begin{pmatrix} a_1(\bar{Z}'') \\ a_2(\bar{Z}'') \end{pmatrix}, \tag{3.44.10}$$

where $\bar{S}''$ is a Stokes constant, which, when t' and t'' are well separated, is approximately equal to $i \exp(2K)$; see (3.7.7b), (3.7.8), (2.3.28) together with (2.3.27b), and (3.44.5). From (3.44.10), (3.44.9) and (3.44.8) we obtain

$$\begin{pmatrix} a_1(\bar{z}'') \\ a_2(\bar{z}'') \end{pmatrix} = \begin{pmatrix} (1 + \bar{S}'\bar{S}'')a_1(z') + \bar{S}''a_2(z') \\ \bar{S}'a_1(z') + a_2(z') \end{pmatrix}, \tag{3.44.11}$$

and from (3.44.1) and (3.44.11) we obtain

$$\psi(\bar{z}'') = [(1 + \bar{S}'\bar{S}'')a_1(z') + \bar{S}''a_2(z')]q^{-1/2}(\bar{z}'') \exp\left[+i \int\limits_{(t')}^{\bar{z}''} q(z)dz\right]$$

$$+ [\bar{S}'a_1(z') + a_2(z')]q^{-1/2}(\bar{z}'') \exp\left[-i \int\limits_{(t')}^{\bar{z}''} q(z)dz\right]. \tag{3.44.12}$$

Noting the location of z'' and $\bar{z}''$ on opposite lips of the cut in Fig. 3.44.1, we realize that

$$q^{-1/2}(\bar{z}'') = -q^{-1/2}(z''), \tag{3.44.13a}$$

$$q(\bar{z}'') = q(z''), \tag{3.44.13b}$$

$$\int\limits_{\bar{t}''}^{\bar{z}''} q(z)dz = \int\limits_{t''}^{z''} q(z)dz. \tag{3.44.13c}$$

Using (3.44.13c) and the definition (3.44.5), we obtain

$$\int\limits_{(t')}^{\bar{z}''} q(z)dz = \int\limits_{(t')}^{(\bar{t}'')} q(z)dz + \int\limits_{(\bar{t}'')}^{\bar{z}''} q(z)dz = -\int\limits_{(t')}^{(t'')} q(z)dz + \int\limits_{(t'')}^{z''} q(z)dz$$

$$= \int\limits_{(t')}^{z''} q(z)dz - 2\int\limits_{(t')}^{(t'')} q(z)dz = \int\limits_{(t')}^{z''} q(z)dz + 2iK. \tag{3.44.14}$$

With the aid of (3.44.13a) and (3.44.14) we can rewrite (3.44.12) into the form

$$\psi(\bar{z}'') = -\exp(-2K)[(1+\bar{S}'\bar{S}'')a_1(z') + \bar{S}''a_2(z')]q^{-1/2}(z'')\exp\left[+i\int_{(t')}^{z''}q(z)dz\right]$$

$$- \exp(2K)[\bar{S}'a_1(z') + a_2(z')]q^{-1/2}(z'')\exp\left[-i\int_{(t')}^{z''}q(z)dz\right]. \quad (3.44.15)$$

The wave function is single-valued when z encircles the transition zeros t' and t'', i.e.,

$$\psi(\bar{z}'') = \psi(z''). \quad (3.44.16)$$

Inserting (3.44.7) and (3.44.15) into (3.44.16), and noting that $a_1(z')$ and $a_2(z')$ can be chosen arbitrarily, we obtain

$$-\exp(-2K)(1+\bar{S}'\bar{S}'') = 1, \quad -\exp(-2K)\bar{S}'' = S',$$
$$-\exp(2K)\bar{S}' = S'', \quad -\exp(2K) = 1 + S'S'',$$

and hence

$$S'S'' = -[\exp(2K)+1], \quad (3.44.17a)$$
$$\bar{S}' = -\exp(-2K)S'', \quad (3.44.17b)$$
$$\bar{S}'' = -\exp(2K)S'. \quad (3.44.17c)$$

Then instead of S' we introduce the quantity ϕ by writing

$$S' = -i[1+\exp(-2K)]^{1/2}\exp(-i\phi) \quad (3.44.18a)$$

and thus obtain from (3.44.17a–c)

$$S'' = -i[1+\exp(-2K)]^{1/2}\exp(2K+i\phi), \quad (3.44.18b)$$
$$\bar{S}' = i[1+\exp(-2K)]^{1/2}\exp(i\phi), \quad (3.44.18c)$$
$$\bar{S}'' = i[1+\exp(-2K)]^{1/2}\exp(2K-i\phi). \quad (3.44.18d)$$

As mentioned below (3.44.2), S' is approximately equal to $-i$, when t' and t'' are well separated, i.e., when $\exp(-2K)$ is sufficiently small. From (3.44.18a) it then follows that ϕ is small compared with unity when the distance between t' and t'' is sufficiently large. When the potential barrier is real and overdense, i.e., when $R(z)$ and $Q^2(z)$ are real on the real z-axis, and hence t' and t'' ($> t'$) lie on the real z-axis, and when furthermore $a_2(z') = a_1^*(z')$, the wave function $\psi(z')$ is real according to (3.44.1). Therefore $\psi(z'')$ must be real, which according to (3.44.7), (3.44.18a,b) and (cf. (3.44.5))

$$\int\limits_{(t')}^{z''} q(z)dz = \int\limits_{(t')}^{(t'')} q(z)dz + \int\limits_{(t'')}^{z''} q(z)dz = \int\limits_{(t'')}^{z''} q(z)dz - iK \qquad (3.44.19)$$

implies that ϕ is real in that particular case.

Inserting (3.44.18a–d) into (3.44.6) and (3.44.11), we get

$$\begin{pmatrix} a_1(z'') \\ a_2(z'') \end{pmatrix} = \begin{pmatrix} 1 & -i[1+\exp(-2K)]^{1/2}\exp(-i\phi) \\ -i[1+\exp(-2K)]^{1/2}\exp(2K+i\phi) & -\exp(2K) \end{pmatrix} \begin{pmatrix} a_1(z') \\ a_2(z') \end{pmatrix}$$

$$(3.44.20)$$

and

$$\begin{pmatrix} a_1(\bar{z}'') \\ a_2(\bar{z}'') \end{pmatrix} = \begin{pmatrix} -\exp(2K) & i[1+\exp(-2K)]^{1/2}\exp(2K-i\phi) \\ i[1+\exp(-2K)]^{1/2}\exp(+i\phi) & 1 \end{pmatrix} \begin{pmatrix} a_1(z') \\ a_2(z') \end{pmatrix},$$

$$(3.44.21)$$

the two-by-two matrices in (3.44.20) and (3.44.21) being $\mathbf{F}(z'', z')$ and $\mathbf{F}\left(\bar{z}'', z'\right)$, respectively.

With the aid of (3.44.20) we can write (3.44.1), with z replaced by z'', as

$$\psi(z'') = \left\{ a_1(z') - i[1+\exp(-2K)]^{1/2}\exp(-i\phi)a_2(z') \right\}$$

$$\times q^{-1/2}(z'')\exp\left[+i\int\limits_{(t')}^{z''} q(z)dz \right]$$

$$- \left\{ i[1+\exp(-2K)]^{1/2}\exp(2K+i\phi)a_1(z') + \exp(2K)a_2(z') \right\}$$

$$\times q^{-1/2}(z'')\exp\left[-i\int\limits_{(t')}^{z''} q(z)dz \right], \qquad (3.44.22)$$

which using (3.44.19) can be written as

$$\psi(z'') = \left\{ \exp(K)a_1(z') - i[\exp(2K)+1]^{1/2}\exp(-i\phi)a_2(z') \right\}$$

$$\times q^{-1/2}(z'')\exp\left[+i\int\limits_{(t'')}^{z''} q(z)dz \right]$$

$$- \left\{ i[\exp(2K)+1]^{1/2}\exp(i\phi)a_1(z') + \exp(K)a_2(z') \right\}$$

$$\times q^{-1/2}(z'')\exp\left[-i\int\limits_{(t'')}^{z''} q(z)dz \right]. \qquad (3.44.23)$$

Even when the absolute value of $\exp(2K)$ is large compared with unity, one cannot in this formula replace $[\exp(2K) + 1]^{1/2}$ by $\exp(K)$ and neglect ϕ, if in higher-order approximation one wants results of great precision. However, if it is sufficient to have a rather approximate formula one can do so, getting

$$\psi(z'') = 2\exp(K - i\pi/4)[a_1(z') - ia_2(z')]$$
$$\times q^{-1/2}(z'')\cos\left[\int_{(t'')}^{z''} q(z)dz + \pi/4\right]. \qquad (3.44.24)$$

This formula cannot, however, be used when $a_1(z')/a_2(z')$ is close to i.

Using (3.44.21), and noting that according to Fig. 3.44.1 and the definition (3.44.5)

$$\int_{(t')}^{\bar{z}''} q(z)dz = \int_{(t')}^{(\bar{t}'')} q(z)dz + \int_{(\bar{t}'')}^{\bar{z}''} q(z)dz$$

$$= -\int_{(t')}^{(t'')} q(z)dz + \int_{(\bar{t}'')}^{\bar{z}''} q(z)dz = \int_{(\bar{t}'')}^{\bar{z}''} q(z)dz + iK, \qquad (3.44.25)$$

we can write (3.44.1), with z replaced by $\bar{z}''$, as

$$\psi(\bar{z}'') = -\{\exp(K)a_1(z') - i[\exp(2K) + 1]^{1/2}\exp(-i\phi)a_2(z')\}$$
$$\times q^{-1/2}(\bar{z}'')\exp\left[+i\int_{(\bar{t}'')}^{\bar{z}''} q(z)dz\right]$$
$$+ \{i[\exp(2K) + 1]^{1/2}\exp(i\phi)a_1(z') + \exp(K)a_2(z')\}$$
$$\times q^{-1/2}(\bar{z}'')\exp\left[-i\int_{(\bar{t}'')}^{\bar{z}''} q(z)dz\right]. \qquad (3.44.26)$$

Comparing (3.44.23) and (3.44.26), we note that because of (3.44.13c) the exponentials in corresponding phase-integral functions are the same, and that the difference in the sign of the right-hand sides is due to (3.44.13a). We can combine (3.44.23) and (3.44.26) into one formula by replacing $\bar{z}''$ by z'' in (3.44.26), deleting the cut in Fig. 3.44.1 from t'' towards infinity, and taking the above-mentioned difference in sign into account. Thus, we get the formula (3.44.23) for $\psi(z'')$ and formula (3.44.26) for $\psi(\bar{z}'')$ summarized into one formula. See Fig. 3.44.2 in which z can move from z' to z'' half of a full turn in either the positive direction or the

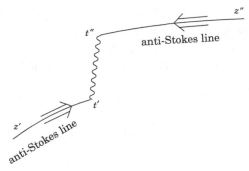

Figure 3.44.2 The double arrows indicate the directions in which $\mathrm{Re}\int^z q\,(z)\,dz$ increases along the anti-Stokes lines. There is a cut (wavy line) joining t' and t''.

negative direction. Similarly to Sections 2.4 and 2.5 we shall denote by A and B, re-spectively, the coefficients of the phase-integral functions representing waves that, on an anti-Stokes line emerging from a transition point t, move away from t and towards t, respectively. Putting thus

$$a_2(z') = A',\tag{3.44.27a}$$

$$a_1(z') = B',\tag{3.44.27b}$$

we have according to (3.44.1) with $z = z'$

$$\psi(z') = A'\,q^{-1/2}(z')\exp\left[-i\int_{(t')}^{z'} q(z)dz\right] + B'\,q^{-1/2}(z')\exp\left[+i\int_{(t')}^{z'} q(z)dz\right],$$

$$\tag{3.44.28}$$

where, as the double arrows in Fig. 3.44.2 indicate, A' and B' are the coefficients of the phase-integral functions that represent waves travelling away from t' and towards t', respectively. Recalling Fig. 3.44.2, we can summarize (3.44.23) and (3.44.26) into the single formula

$$\psi(z'') = A''\,q^{-1/2}(z'')\exp\left[-i\int_{(t'')}^{z''} q(z)dz\right] + B''\,q^{-1/2}(z'')\exp\left[+i\int_{(t'')}^{z''} q(z)dz\right],$$

$$\tag{3.44.29}$$

where A'', which is equal to $a_2(z'')$ or $a_2(\bar{z}'')$, and B'', which is equal to $a_1(z'')$ or $a_1(\bar{z}'')$, are coefficients of the phase-integral functions that represent waves travel-ling away from t'' and towards t'', respectively, and where

$$\begin{pmatrix} A'' \\ B'' \end{pmatrix} = \pm \begin{pmatrix} \exp(K) & i[\exp(2K)+1]^{1/2}\exp(i\phi) \\ i[\exp(2K)+1]^{1/2}\exp(-i\phi) & -\exp(K) \end{pmatrix}\begin{pmatrix} A' \\ B' \end{pmatrix},$$

$$\tag{3.44.30}$$

the upper sign applying when one traces $\psi(z)$ from z' to z'' by moving in the positive direction, and the lower sign applying when one traces $\psi(z)$ from z' to z'' by moving in the negative direction. The reason for the double sign in (3.44.30) is that $q^{-1/2}(z)$ changes sign when z moves one full turn around t' and t'', as is displayed in (3.44.13a). The phase of $Q(z)$ (or of $q(z)$) is needed only to recognize the directions of propagation of the waves represented by the phase-integral functions.

The inverse of the matrix in (3.44.30) is

$$\left[\pm \begin{pmatrix} \exp(K) & i[\exp(2K)+1]^{1/2}\exp(i\phi) \\ i[\exp(2K)+1]^{1/2}\exp(-i\phi) & -\exp(K) \end{pmatrix} \right]^{-1}$$

$$= \mp \begin{pmatrix} \exp(K) & i\left[1\exp(2K)+1\right]^{1/2}\exp(i\phi) \\ i[\exp(2K)+1]^{1/2}\exp(-i\phi) & -\exp(K) \end{pmatrix}.$$

$$(3.44.31)$$

Note that the only difference between the matrices in (3.44.30) and (3.44.31) is that they have opposite signs, which is due to the fact that if one moves from z' to z'' in the positive direction, one moves along the same path from z'' to z' in the negative direction.

The quantity $[\exp(2K)+1]^{1/2}$ appears in (3.44.30) as a consequence of the fact that the wave function $\psi(z)$ is single-valued when z moves a full turn around the transition zeros t' and t'' in Fig. 3.44.2. A treatment of the problem using the Stokes constants for two well-separated transition zeros, obtained in Section 3.7, would amount to the replacement of $[1+\exp(-2K)]^{1/2}$ by unity and ϕ by zero in the expressions (3.44.18a–d) for the Stokes constants. The phase-integral wave function $\psi(z)$ is then not single-valued when z moves a full turn around t' and t'', and thus one does not get the same wave function when z moves from z' to z'' in the positive direction as when z moves from z' to z'' in the negative direction.

The situation is different when t' and t'' are joined by an anti-Stokes line; such a cluster of two transition zeros can support a localized state. When one then uses the simple Stokes constants for two well-separated transition zeros, the wave function is, according to Section 3.7, single-valued when z moves a full turn around each transition zero and hence also when z moves a full turn around the cluster consisting of both transition zeros. The encircling of the two transition zeros involves in this case six Stokes constants, while the encircling of the two transition zeros delimiting a complex potential barrier involves only four Stokes constants, since there are only four sectors delimited by anti-Stokes lines; see Fig. 3.44.1. Since for a complex barrier there is no anti-Stokes line joining t' and t'', z cannot move a full turn around the barrier by jumping between and moving along anti-Stokes lines in such a way that z moves first a full turn around one of the transition zeros and then a full turn around the other transition zero.

By means of the comparison equation technique Fröman, Fröman and Lundborg (1996) derived the following approximate, but very accurate formula for ϕ in the $(2N + 1)$th order of the phase-integral approximation (see their eqs. (5.5.30), (5.5.25a–g), (5.4.23) and (5.4.21) with $\lambda = 1$)

$$\phi = \frac{1}{2i} \ln \frac{\Gamma(1/2 + i\bar{K})}{\Gamma(1/2 - i\bar{K})} - \bar{K} \ln \bar{K}_0 + \sum_{n=0}^{N} \phi^{(2n+1)}, \qquad (3.44.32a)$$

$$\bar{K} = K/\pi, \qquad (3.44.32b)$$

where $\bar{K}_0$ is $\bar{K}$ in the first-order approximation, and the first three quantities $\phi^{(2n+1)}$ are given by (2.5.13a–c) together with (2.5.12b). Note that ϕ in (3.44.32) is associated with a complex barrier and is complex, while $\tilde{\phi}$ in (2.5.11) is associated with a real barrier and is real. Although (3.44.32a) is also valid when the barrier is thick, the following simpler formula (obtained from eq. (5.5.31) with $\lambda = 1$ in Fröman, Fröman and Lundborg (1996)) is useful for this case (cf. (2.5.14))

$$\phi = -\phi^{(2N+3)}, \quad \text{thick barrier,} \quad (2N+1)\text{th-order approximation.} \qquad (3.44.33)$$

Alternative choice of phase of base function. We shall also give formulas corresponding to an alternative choice of phase of the base function. In this case, which is illustrated in Fig. 3.44.3, we shall put a bar on the base function, the phase integrand and the coefficients A and B. We can transform the formulas already derived to conform to this new situation by putting $Q(z) = -\bar{Q}(z)$. As a consequence of this we have

$$q(z) = -\bar{q}(z), \qquad (3.44.34a)$$

$$q^{-1/2}(z) = \alpha \bar{q}^{-1/2}(z), \qquad (3.44.34b)$$

where

$$\alpha^2 = -1. \qquad (3.44.35)$$

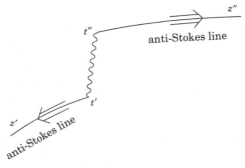

Figure 3.44.3 The double arrows indicate the directions in which $\mathrm{Re} \int^z \bar{q}(z)\,dz$ increases along the anti-Stokes lines. As in Fig. 3.44.2 there is a cut (wavy line) joining t' and t''.

Using (3.44.34a,b) and defining

$$\bar{A}' = \alpha A', \qquad (3.44.36a)$$

$$\bar{B}' = \alpha B', \qquad (3.44.36b)$$

$$\bar{A}'' = \alpha A'', \qquad (3.44.36c)$$

$$\bar{B}'' = \alpha B'', \qquad (3.44.36d)$$

we obtain from (3.44.28)

$$\psi(z') = \bar{A}'\bar{q}^{-1/2}(z')\exp\left[+i\int\limits_{(t')}^{z'}\bar{q}(z)dz\right] + \bar{B}'\bar{q}^{-1/2}(z')\exp\left[-i\int\limits_{(t')}^{z'}\bar{q}(z)dz\right],$$

$$(3.44.37)$$

and from (3.44.29)

$$\psi(z'') = \bar{A}''\bar{q}^{-1/2}(z'')\exp\left[+i\int\limits_{(t'')}^{z''}\bar{q}(z)dz\right] + \bar{B}''\bar{q}^{-1/2}(z'')\exp\left[-i\int\limits_{(t'')}^{z''}\bar{q}(z)dz\right].$$

$$(3.44.38)$$

With the aid of (3.44.36a–d) we can rewrite (3.44.30) into the form

$$\begin{pmatrix}\bar{A}''\\ \bar{B}''\end{pmatrix} = \pm \begin{pmatrix} \exp(K) & i[\exp(2K)+1]^{1/2}\exp(i\phi) \\ i[\exp(2K)+1]^{1/2}\exp(-i\phi) & -\exp(K) \end{pmatrix}\begin{pmatrix}\bar{A}'\\ \bar{B}'\end{pmatrix},$$

$$(3.44.39)$$

and with the aid of (3.44.34a) the original definition (3.44.5) of K can be written as

$$K = -i\int\limits_{(t')}^{(t'')}\bar{q}(z)dz = -\frac{1}{2i}\int\limits_{\Lambda}\bar{q}(z)dz, \qquad (3.44.40)$$

where, as in (3.44.5), Λ is a closed contour, encircling t' and t'', along which the integration is performed in the positive direction.

3.45 Connection formula for a real, single-hump potential barrier

Obtain the connection formula for a real, single-hump potential barrier by particularization of the connection formula for a complex barrier obtained in Section 3.44.

Solution. Since we are considering points on the real z-axis, we replace z' by x' and z'' by x'' in Section 3.44. Referring to the case of the overdense barrier in

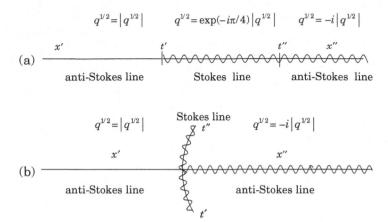

Figure 3.45.1 (a) Subbarrier penetration, and (b) superbarrier transmission. The wavy lines denote cuts. The cut between t' and t'' coincides with the Stokes line joining t' and t''.

Fig. 3.45.1(a), we note that

$$\int_{(t')}^{x'} q(z)dz = \operatorname{Re} \int_{(t')}^{x'} q(z)dz, \tag{3.45.1a}$$

$$\int_{(t'')}^{x''} q(z)dz = \operatorname{Re} \int_{(t'')}^{x''} q(z)dz. \tag{3.45.1b}$$

Referring to the case of the underdense barrier in Fig. 3.45.1(b), and using (3.44.5), we obtain

$$\int_{(t')}^{x'} q(z)dz = \operatorname{Re} \int_{(t')}^{x'} q(z)dz + i\operatorname{Im} \int_{(t')}^{x'} q(z)dz = \operatorname{Re} \int_{(t')}^{x'} q(z)dz + \tfrac{1}{2} \int_{(t')}^{(t'')} q(z)dz$$

$$= \operatorname{Re} \int_{(t')}^{x'} q(z)dz - \tfrac{1}{4} \int_{\Lambda} q(z)dz = \operatorname{Re} \int_{(t')}^{x'} q(z)dz - \tfrac{1}{2}iK, \tag{3.45.2a}$$

$$\int_{(t'')}^{x''} q(z)dz = \operatorname{Re} \int_{(t'')}^{x''} q(z)dz + i\operatorname{Im} \int_{(t'')}^{x''} q(z)dz = \operatorname{Re} \int_{(t'')}^{x''} q(z)dz - \tfrac{1}{2} \int_{(t')}^{(t'')} q(z)dz$$

$$= \operatorname{Re} \int_{(t'')}^{x''} q(z)dz + \tfrac{1}{4} \int_{\Lambda} q(z)dz = \operatorname{Re} \int_{(t'')}^{x''} q(z)dz + \tfrac{1}{2}iK, \tag{3.45.2b}$$

where the integrals with the 'limits of integration' (t') and (t'') are performed to the left of the cut, and Λ, as in (3.44.5), is a closed contour of integration, encircling t'

and t'', along which the integration is performed in the positive direction, and the cut along the real axis in Fig. 3.45.1(b) is disregarded. The formulas (3.45.1a,b) and (3.45.2a,b) can be summarized as

$$\int_{(t')}^{x'} q(z)dz = \text{Re} \int_{(t')}^{x'} q(z)dz - \tfrac{1}{2}ip, \tag{3.45.3a}$$

$$\int_{(t'')}^{x''} q(z)dz = \text{Re} \int_{(t'')}^{x''} q(z)dz + \tfrac{1}{2}ip, \tag{3.45.3b}$$

where

$$p = \begin{cases} 0 & \text{for the case in Fig. 3.45.1(a),} \\ K & \text{for the case in Fig. 3.45.1(b).} \end{cases} \tag{3.45.4}$$

Using (3.45.3a,b), we can write (3.44.28) and (3.44.29) as

$$\psi(x') = \tilde{A}'|q^{-1/2}(x')|\exp\left[+i\left|\text{Re}\int_{(t')}^{x'} q(z)dz\right|\right]$$

$$+ \tilde{B}'|q^{-1/2}(x')|\exp\left[-i\left|\text{Re}\int_{(t')}^{x'} q(z)dz\right|\right], \tag{3.45.5a}$$

$$\psi(x'') = \tilde{A}''|q^{-1/2}(x'')|\exp\left[+i\left|\text{Re}\int_{(t'')}^{x''} q(z)dz\right|\right]$$

$$+ \tilde{B}''|q^{-1/2}(x'')|\exp\left[-i\left|\text{Re}\int_{(t'')}^{x''} q(z)dz\right|\right], \tag{3.45.5b}$$

where

$$\tilde{A}' = A'\exp(-p/2), \tag{3.45.6a}$$
$$\tilde{B}' = B'\exp(+p/2), \tag{3.45.6b}$$
$$\tilde{A}'' = iA''\exp(-p/2), \tag{3.45.7a}$$
$$\tilde{B}'' = iB''\exp(+p/2). \tag{3.45.7b}$$

From Fig. 3.45.1 it is seen that a path from x' to x'' in the upper half of the complex z-plane proceeds in the negative direction. Therefore we obtain from (3.44.30) with the lower sign, (3.45.6a,b) and (3.45.7a,b)

$$\begin{pmatrix} \tilde{A}'' \\ \tilde{B}'' \end{pmatrix} = \tilde{\mathbf{M}} \begin{pmatrix} \tilde{A}' \\ \tilde{B}' \end{pmatrix}, \tag{3.45.8}$$

where

$$\tilde{\mathbf{M}} = \begin{pmatrix} \exp(K - i\pi/2) & [\exp(2K) + 1]^{1/2}\exp(+i\tilde{\phi}) \\ [\exp(2K) + 1]^{1/2}\exp(-i\tilde{\phi}) & \exp(K + i\pi/2) \end{pmatrix}, \tag{3.45.9a}$$

$$\det \tilde{\mathbf{M}} = -1. \tag{3.45.9b}$$

Here

$$\tilde{\phi} = \phi + ip, \tag{3.45.10}$$

with ϕ given by (3.44.32a), and (cf. (3.44.5) and (3.44.32b))

$$K = \pi\bar{K} = \frac{1}{2i} \int_{\Lambda} q(z)dz, \tag{3.45.11}$$

where Λ, as already mentioned, is a closed contour, encircling t' and t'', along which the integration is performed in the positive direction. With due regard to the phase of $q(z)$, obtained according to Fig. 3.45.1, we note that in the first-order approximation K is positive in the subbarrier case but negative in the superbarrier case. Introducing into formula (2.5.6a), which is in principle exact, the approximations (2.5.10a,b), i.e., $\theta \approx \exp(K)$ and $\vartheta \approx 0$, one obtains (3.45.9a), but it remains to show that the expressions (2.5.11) and (3.45.10) for $\tilde{\phi}$ agree, and this will now be done.

From (3.45.10) and (3.45.4) we obtain

$$\tilde{\phi} = \begin{cases} \phi & \text{when } K_0 > 0, \quad \text{Fig. 3.45.1(a)}, \\ \phi + iK & \text{when } K_0 < 0, \quad \text{Fig. 3.45.1(b)}, \end{cases} \tag{3.45.12}$$

where K_0 is K in the first-order approximation. When $t'' - t'$ rotates in the positive direction from the situation in Fig. 3.45.1(a) to the situation in Fig. 3.45.1(b), the argument of $t'' - t'$ increases from 0 to $\pi/2$ and the argument of $\bar{K}_0$ increases from 0 to π, and $\ln \bar{K}_0$ thus changes from $\ln |\bar{K}_0|$ to $\ln |\bar{K}_0| + i\pi$. Therefore, in the subbarrier case illustrated in Fig. 3.45.1(a) we have $-\bar{K} \ln \bar{K}_0 = -\bar{K} \ln |\bar{K}_0|$, while in the superbarrier case illustrated in Fig. 3.45.1(b) we have instead $-\bar{K} \ln \bar{K}_0 = -\bar{K}(\ln |\bar{K}_0| + i\pi) = -\bar{K} \ln |\bar{K}_0| - iK$, since $\bar{K} = K/\pi$ according to (3.45.11). From (3.44.32a) and (3.45.12) we therefore obtain, when we recall that $\bar{K}$ is real for a real potential barrier,

$$\tilde{\phi} = \arg \Gamma(1/2 + i\bar{K}) - \bar{K} \ln |\bar{K}_0| + \sum_{n=0}^{N} \phi^{(2n+1)}, \tag{3.45.13}$$

in agreement with (2.5.11). The first three of the quantities $\phi^{(2n+1)}$ are given by (2.5.13a–c). This result can be found using different notation in Fröman, Fröman, Myhrman and Paulsson (1972).

From (3.45.9a,b) we obtain (cf. (2.5.6a) and (2.5.7))

$$\tilde{\mathbf{M}}^{-1} = \tilde{\mathbf{M}}, \qquad (3.45.14)$$

and hence (3.45.8) can be written in the alternative form

$$\begin{pmatrix} \tilde{A}' \\ \tilde{B}' \end{pmatrix} = \tilde{\mathbf{M}} \begin{pmatrix} \tilde{A}'' \\ \tilde{B}'' \end{pmatrix}. \qquad (3.45.15)$$

From (3.45.8) and (3.45.15) together with (3.45.9a) it is seen that the quantity $\tilde{\phi}$ is the same whether one considers the connection from left to right or from right to left.

For a thick barrier one can often neglect $\tilde{\phi}$ in (3.45.9a), but if one wants to obtain a very high accuracy using a higher-order approximation, one has also to retain $\tilde{\phi}$ in (3.45.9a) when the barrier is thick.

When the barrier is so thick that $\exp(-2K) \ll 1$, we have approximately $[\exp(2K) + 1]^{1/2} \approx \exp(K) + \frac{1}{2}\exp(-K)$. Introducing this approximation into (3.45.9a) with $\tilde{\phi}$ neglected, we obtain

$$\tilde{\mathbf{M}} = \begin{pmatrix} -i\exp(K) & \exp(K) + \frac{1}{2}\exp(-K) \\ \exp(K) + \frac{1}{2}\exp(-K) & i\exp(K) \end{pmatrix}. \qquad (3.45.16)$$

If one treats the penetration problem for a thick, single-hump potential barrier *erroneously* by using the connection formulas associated with well-separated turning points, but disregarding their one-directional nature, i.e., if one treats the penetration problem by means of the parameterized formulas (2.4.23), (2.4.26), (2.4.28a,b) and (2.4.29a,b) with the approximate values $\alpha = 1$ and $\beta = 0$ and the erroneous value $\gamma = 0$, one obtains for the matrix $\tilde{\mathbf{M}}$ the erroneous formula

$$\tilde{\mathbf{M}} = \begin{pmatrix} -i[\exp(K) - \frac{1}{4}\exp(-K)] & \exp(K) + \frac{1}{4}\exp(-K) \\ \exp(K) + \frac{1}{4}\exp(-K) & i[\exp(K) - \frac{1}{4}\exp(-K)] \end{pmatrix}, \; \textit{erroneous formula,}$$

$$(3.45.17)$$

which differs from the corresponding correct approximate formula (3.45.16) as regards the coefficients of $\exp(-K)$. This has been verified in a concrete way for a purely parabolic barrier and the first-order approximation by Fröman and Fröman (1965, p. 51).

3.46 Energy levels of a particle in a smooth double-well potential, when no symmetry requirement is imposed

Determine the energy levels of a quantal particle in an unspecified, smooth, double-well potential $V(z)$, i.e., a potential consisting of two wells separated by a barrier; see Fig. 3.46.1.

Solution. With the aid of a connection formula we shall first trace a bound-state wave function from the classically forbidden region to the left of t_0'' into the left-hand potential well. With the aid of the connection formula for a real potential barrier, we shall then trace the wave function across the barrier into the right-hand potential well. There it can be fitted to the expression for the wave function obtained with the aid of the connection formula for tracing a bound-state wave function from the classically forbidden region to the right of t_2' into the right-hand potential well. The fitting yields the quantization condition for the calculation of the energy levels. If the double-well potential is symmetric, and the potential barrier is thick, there are pairs of close-lying energy levels, and in the formulas obtained as just described the errors in the positions of the pairs are much larger than the splittings of the levels in each pair. In order to demonstrate this fact we shall in the solution below use formulas from Sections 2.4 and 2.5, which are in principle exact. In the formulas obtained in this way we shall then omit small quantities that cannot be determined within the framework of the phase-integral method, but for which one has upper limits.

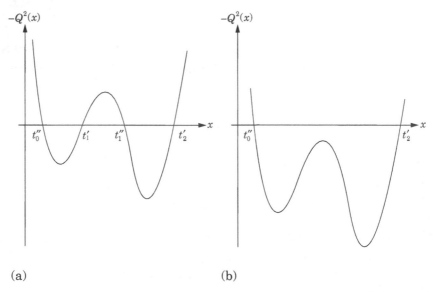

(a)   (b)

Figure 3.46.1 Double-well potential for which the transition zeros t_1' and t_1'' associated with the barrier are real in (a) but complex conjugate in (b).

If the classically forbidden region to the left of t_0'' is infinitely thick, i.e., if the phase integral from the turning point to the point at which the boundary condition is imposed, is infinitely large, the bound-state wave function to the left of t_0'' is

$$\psi(x) = C_0''|q^{-1/2}(x)|\exp\left[+\left|\int_{(t_0'')}^{x} q(z)dz\right|\right] + D_0''|q^{-1/2}(x)|\exp\left[-\left|\int_{(t_0'')}^{x} q(z)dz\right|\right]$$

$$\approx D_0''|q^{-1/2}(x)|\exp\left[-\left|\int_{(t_0'')}^{x} q(z)dz\right|\right], \quad x < t_0'', \tag{3.46.1}$$

where $(C_0''/D_0'') \exp[2|\int_{(t_0'')}^{x} q(z)dz|]$ tends to zero as x tends to the point at which the boundary condition that the wave function be equal to zero is imposed. We assume D_0'' to be real and positive. In the left-hand potential well the wave function, obtained from (3.46.1) with the aid of (2.4.23), (2.4.26), (2.4.28a,b) and (2.4.8c), is

$$\psi(x) = 2\alpha_0'' D_0''|q^{-1/2}(x)|\cos\left[\left|\int_{(t_0'')}^{x} q(z)dz\right| - \beta_0'' - \pi/4\right]. \tag{3.46.2}$$

This wave function can also be written as

$$\psi(x) = \Omega'|q^{-1/2}(x)|\cos\left[\left|\operatorname{Re}\int_{(t_1')}^{x} q(z)dz\right| + \delta' - \pi/4\right], \tag{3.46.3}$$

with

$$\Omega' = 2\alpha_0'' D_0''(> 0), \tag{3.46.4a}$$

$$\delta' = \pi/2 - L_1 + \beta_0'', \tag{3.46.4b}$$

where

$$L_1 = \left|\operatorname{Re}\int_{(t_0'')}^{(t_1')} q(z)dz\right|. \tag{3.46.5}$$

In the right-hand potential well the same wave function is, according to (2.5.19a,b) and (2.5.23a,b),

$$\psi(x) = \Omega''|q^{-1/2}(x)|\cos\left[\left|\operatorname{Re}\int_{(t_1'')}^{x} q(z)dz\right| + \delta'' - \pi/4\right], \tag{3.46.6}$$

with

$$\delta'' = \arctan\left[\frac{(\theta^2+1)^{1/2}-\theta}{(\theta^2+1)^{1/2}+\theta}\tan(\pi/2+\tilde{\phi}/2+\vartheta/2-\delta')\right]+\tilde{\phi}/2-\vartheta/2,$$

(3.46.7a)

$$\Omega'' = \Omega'\left\{\left[(\theta^2+1)^{1/2}-\theta\right]^2+4\theta(\theta^2+1)^{1/2}\cos^2(\pi/2+\tilde{\phi}/2+\vartheta/2-\delta')\right\}^{1/2}(>0),$$

(3.46.7b)

the angle $\delta'' - \tilde{\phi}/2 + \vartheta/2$ lying in the same quadrant as $\pi/2 + \tilde{\phi}/2 + \vartheta/2 - \delta'$ (mod 2π). According to (2.5.10a,b)

$$\theta \approx \exp(K),$$

(3.46.8a)

$$\vartheta \approx 0,$$

(3.46.8b)

where from (2.5.2)

$$K = \frac{1}{2i}\int_{\Lambda} q(z)dz,$$

(3.46.9)

Λ being a closed contour, encircling t_1' and t_1'', along which the integration is performed in the direction that in the first-order approximation makes K positive for a superdense barrier but negative for an underdense barrier. We rewrite (3.46.6) in the form

$$\psi(x) = \Omega''|q^{-1/2}(x)|\cos\left[\left|\mathrm{Re}\int_{(t_2')}^{x}q(z)dz\right|+\pi/2-L_2-\delta''-\pi/4\right],$$

(3.46.10)

where

$$L_2 = \left|\mathrm{Re}\int_{(t_1'')}^{(t_2')}q(z)dz\right|.$$

(3.46.11)

If the classically forbidden region to the right of t_2' is infinitely thick, i.e., if the phase integral from the turning point to the point at which the boundary condition is imposed is infinitely large, the bound-state wave function to the right of t_2' is

$$\psi(x) = C_2'|q^{-1/2}(x)|\exp\left[+\left|\int_{(t_2')}^{x}q(z)dz\right|\right]+D_2'|q^{-1/2}(x)|\exp\left[-\left|\int_{(t_2')}^{x}q(z)dz\right|\right]$$

$$\approx D_2'|q^{-1/2}(x)|\exp\left[-\left|\int_{(t_2')}^{x}q(z)dz\right|\right],\quad x>t_2',$$

(3.46.12)

where $(C_2'/D_2')\exp[2|\int_{(t_2')}^x q(z)dz|]$ tends to zero as x tends to the point at which the boundary condition that the wave function be equal to zero is imposed. In the right-hand potential well the wave function, obtained from (3.46.12) with the aid of (2.4.23), (2.4.26), (2.4.28a,b) and (2.4.8c), is

$$\psi(x) = 2\alpha_2' D_2' |q^{-1/2}(x)| \cos\left[\left|\int_{(t_2')}^x q(z)dz\right| - \beta_2' - \pi/4\right]. \qquad (3.46.13)$$

By identifying (3.46.10) and (3.46.13) we obtain

$$\Omega'' = 2\alpha_2' D_2', \qquad (3.46.14a)$$
$$\delta'' = \pi/2 - L_2 + \beta_2'. \qquad (3.46.14b)$$

Using (3.46.4a), (3.46.14a), (3.46.7b) and (3.46.4b), we get

$$\frac{D_2'}{D_0''} = \frac{\alpha_0''}{\alpha_2'}\{[(\theta^2+1)^{1/2} - \theta]^2 + 4\theta(\theta^2+1)^{1/2}$$
$$\times \sin^2(L_1 - \pi/2 + \tilde{\phi}/2 + \vartheta/2 - \beta_0'')\}^{1/2}. \qquad (3.46.15)$$

Writing (3.46.7a) as

$$\tan(\delta' - \tilde{\phi}/2 - \vartheta/2)\tan(\delta'' - \tilde{\phi}/2 + \vartheta/2) = \frac{(\theta^2+1)^{1/2} - \theta}{(\theta^2+1)^{1/2} + \theta} \qquad (3.46.16)$$

and using (3.46.4b) and (3.46.14b), we obtain

$$\tan(L_1 - \pi/2 + \tilde{\phi}/2 + \vartheta/2 - \beta_0'')\tan(L_2 - \pi/2 + \tilde{\phi}/2 - \vartheta/2 - \beta_2') = \frac{(\theta^2+1)^{1/2} - \theta}{(\theta^2+1)^{1/2} + \theta}.$$
$$(3.46.17)$$

This formula can also be written in the form

$$\frac{\cos(L_1 - L_2 + \vartheta + \beta_2' - \beta_0'')}{\cos(L_1 + L_2 - \pi + \tilde{\phi} - \beta_2' - \beta_0'')} = (1 + 1/\theta^2)^{1/2}, \qquad (3.46.18)$$

where the arguments of the cosines in the numerator and in the denominator are the difference and the sum, respectively, of the arguments of tan in (3.46.17).

Using (3.46.8a,b) and (2.4.8a,b) we obtain from (3.46.15), (3.46.17) and (3.46.18), respectively, the approximate formulas

$$\frac{D_2'}{D_0''} = (\{[\exp(2K)+1]^{1/2} - \exp(K)\}^2$$
$$+ 4\exp(K)[\exp(2K)+1]^{1/2}\sin^2(L_1 - \pi/2 + \tilde{\phi}/2))^{1/2}, \qquad (3.46.15')$$

$$\tan(L_1 - \pi/2 + \tilde{\phi}/2)\tan(L_2 - \pi/2 + \tilde{\phi}/2) = \frac{[\exp(2K)+1]^{1/2} - \exp(K)}{[\exp(2K)+1]^{1/2} + \exp(K)},$$
$$(3.46.17')$$

$$\frac{\cos(L_1 - L_2)}{\cos(L_1 + L_2 - \pi + \tilde{\phi})} = [1 + \exp(-2K)]^{1/2}. \qquad (3.46.18')$$

The quantization condition (3.46.17′) is seen to be the same as the quantization condition (40) in N. Fröman (1966c), when one remembers that σ in that paper is the same as $-\tilde{\phi}/2$ in the present book.

3.47 Energy levels of a particle in a smooth, *symmetric*, double-well potential

Find the energy levels of a quantal particle in a smooth, symmetric, double-well potential with a barrier of arbitrary thickness by particularizing the quantization condition derived in Section 3.46 and, alternatively, by using the fact that the wave function is either symmetric or anti-symmetric with respect to the centre of the barrier. Then obtain an expression for the splitting of the close-lying energy levels in terms of a frequency of oscillation using formulas in Section 3.26.

Solution obtained by particularization of (3.46.18). When the double-well potential is symmetric, we put in the problem in Section 3.46 $L_1 = L_2 = L$, $\vartheta = 0$ according to (2.5.8), $\beta_2' = \beta_0''$ by a similar argumentation as in the addendum concerning the parameters α, β and γ in Section 3.22, and $\theta \approx \exp(K)$ according to (2.5.10a). The quantization condition (3.46.18) then becomes

$$2(L - \pi/2 + \tilde{\phi}/2 - \beta_0'') = 2\pi s \mp \arccos[1 + \exp(-2K)]^{-1/2}$$
$$= 2\pi s \mp \arctan\exp(-K),$$

where s is an integer, i.e.,

$$L = (s + 1/2)\pi - \tfrac{1}{2}\tilde{\phi} + \beta_0'' \mp \tfrac{1}{2}\arctan\exp(-K). \qquad (3.47.1)$$

Here the upper sign applies when the wave function is symmetric, and the lower sign when it is anti-symmetric. This will be shown later in this solution.

Well below the top of the barrier of a symmetric double-oscillator the energy levels occur in close-lying pairs, the splittings of which are obtained from (3.47.1) with small relative errors, while the positions of the pairs are obtained with errors that are in general large compared with the splittings when β_0'' is neglected, unless one uses a phase-integral approximation of conveniently high order. The reason for this is further explained below (3.47.5).

The structure of close-lying levels fades as the energy approaches and goes over the top of the barrier. At high superbarrier energies the quantization condition of the double-oscillator degenerates to a single-well quantization condition, as has been verified for a general (not necessarily symmetric) double-oscillator by N. Fröman (1966c); see in particular eq. (39) there.

If one has from the beginning a symmetric single-well potential and changes it to a symmetric double-well potential by introducing a symmetric barrier, one has from the beginning well-separated energy levels of alternating even and odd parity. As the barrier height increases continuously, neighbouring energy levels of even and odd parity approach each other, and, when the barrier finally is infinitely high, the two energy levels in each pair coincide and form a doubly degenerate state.

Alternative solution. We begin in the same way as in Section 3.46 and obtain according to (3.46.3), (3.46.4b) and (3.46.5)

$$\psi(x) = \Omega' \left| q^{-1/2}(x) \right| \cos \left[\left| \operatorname{Re} \int_{(t'_1)}^{x} q(z)dz \right| - L + \beta''_0 + \pi/4 \right], \quad (3.47.2)$$

where t'_1 is the left-hand transition zero associated with the barrier in Fig. 3.46.1, now assumed to be symmetric, and

$$L = \left| \operatorname{Re} \int_{(t''_0)}^{(t'_1)} q(z)dz \right| = \left| \operatorname{Re} \int_{(t''_1)}^{(t'_2)} q(z)dz \right|. \quad (3.47.3)$$

The bound-state wave functions associated with a symmetric double-well potential are either symmetric or anti-symmetric with respect to the centre of the barrier. The logarithmic derivative of the wave function at the centre of the barrier is therefore equal to either zero or infinity; and the wave function in the classically allowed region to the left of the barrier is according to (2.5.19a) together with (3.22.5) and (3.22.6), which are valid even when the barrier is thin, and (2.5.10a)

$$\psi(x) = \pm\Omega' \left| q^{-1/2}(x) \right| \cos \left[\left| \operatorname{Re} \int_{(t'_1)}^{x} q(z)dz \right| + \tfrac{1}{2}\tilde{\phi} \pm \tfrac{1}{2}\arctan\exp(-K) - \pi/4 \right],$$

$$(3.47.4)$$

the upper and the lower sign in the argument of the cosine being used depending on whether the wave function is symmetric or anti-symmetric.

Identifying (3.47.2) and (3.47.4), we obtain in agreement with (3.47.1)

$$L = (s + 1/2)\pi - \tfrac{1}{2}\tilde{\phi} + \beta''_0 \mp \tfrac{1}{2}\arctan\exp(-K), \quad (3.47.5)$$

where, as in (3.47.4), the upper sign applies when the wave function is symmetric, and the lower sign applies when it is anti-symmetric. According to (2.4.8b) β''_0 is of the magnitude $O(\mu)$. For a thick barrier $\tilde{\phi}$ is also small compared with unity according to (2.5.14) and (2.5.13b,c), and β''_0 as well as $\tilde{\phi}$ can be expected to assume very nearly the same values for a pair of close-lying energy levels. It is therefore seen from (3.47.5) that the splitting $\Delta L = \arctan \exp(-K)$ is very accurate, while the positions of the separate levels in the pair are in general displaced by distances much larger than the splitting, when β_0'' is neglected; see (2.4.6) with $i w(x_1) = K/2$.

Splitting of the energy levels. Assuming the barrier to be thick, we shall express the splitting ΔE of the two energy levels corresponding to the quantum number s in terms of an oscillation frequency by means of formulas obtained in Section 3.26. To do so we let (3.47.5) with the lower and the upper sign, respectively, correspond to (3.26.1) and (3.26.2) with

$$(\Delta + \Delta_1)\pi = \tfrac{1}{2}\pi - \tfrac{1}{2}\tilde{\phi} + \beta_0'' + \tfrac{1}{2}\exp(-K) \qquad (3.47.6a)$$

and

$$(\Delta + \Delta_2)\pi = \tfrac{1}{2}\pi - \tfrac{1}{2}\tilde{\phi} + \beta_0'' - \tfrac{1}{2}\exp(-K), \qquad (3.47.6b)$$

respectively, where we have replaced $\arctan \exp(-K)$ by $\exp(-K)$. From (3.26.9) and (3.47.6a,b) we then obtain the splitting of the energy levels as

$$\Delta E = E_1 - E_2 = h\nu(\Delta_1 - \Delta_2) = h\nu \exp(-K)/\pi, \qquad (3.47.7)$$

where ν is the frequency of oscillation for a classical particle, with the energy obtained from (3.26.4) with $\Delta = 1/2$, in one of the wells, if $R(z) = Q^2(z)$. Recalling the remark at the end of Section 3.27, we see from (3.47.7) that the splitting ΔE is equal to $\exp(-K)/\pi$ times the distance between the energy levels corresponding to the quantum numbers s and $s + 1$.

3.48 Determination of the quasi-stationary energy levels of a particle in a radial potential with a thick single-hump barrier

Determine the quasi-stationary energy levels of a quantal particle in a radial potential with a single-hump barrier, which is so thick that the position and the width of a narrow level can be represented by the real and the imaginary part, respectively, of a complex energy.

Solution. We denote the radial variable by z, since part of the results can also be used in non-radial problems. The energy E has a small, negative, imaginary part, and therefore the transition zeros t_0'', t_1' and t_1'' in Fig. 3.48.1 do not lie exactly on the real z-axis. The physical potential is $V(z)$, the effective potential is $V(z) + \hbar^2 l(l + 1)/(2mz^2)$, and we choose

$$Q^2(z) = (2m/\hbar^2)[E - V(z)] - (l + 1/2)^2/z^2. \qquad (3.48.1)$$

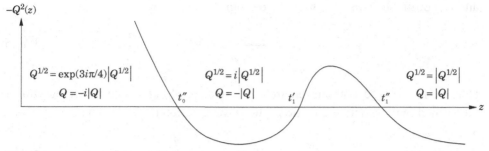

Figure 3.48.1 Except for an additive constant and a multiplicative factor, $-Q^2(z)$ is a radial potential with a single-hump barrier. Restricting z to move in the upper half of the complex z-plane, we indicate in the figure the values on the real z-axis of the phases of Q and $Q^{1/2}$. These are only approximate, since the energy contains a small, negative imaginary part.

The boundary condition that $\psi(z)$ be a purely outgoing wave for large positive values of z requires great care. In Section 2.3.7 we explained how this boundary condition on the real z-axis can be replaced by the corresponding boundary condition on the anti-Stokes line that in Fig. 3.48.1 emerges from t_1'' and for large positive values of z proceeds in the upper half of the complex z-plane in a direction almost parallel to the positive real z-axis; it coincides with part of the real axis when E is real. On this anti-Stokes line the wave function is

$$\psi(z) = A_1'' q^{-1/2}(z)\exp\left[+i\int\limits_{(t_1'')}^{z} q(z)dz\right]. \qquad (3.48.2)$$

We shall now obtain an expression for the same wave function on the anti-Stokes line that emerges from t_1' and lies close to the real z-axis (and coincides with part of the real z-axis when E is real). The result obtained by means of the connection formula for a complex barrier, given by (3.44.37), (3.44.38), (3.44.39) with the lower sign in (3.44.39) and with the bars omitted, and (3.44.31) with the lower signs, is

$$\psi(z) = A_1' q^{1/2}(z)\exp\left[+i\int\limits_{(t_1')}^{z} q(z)dz\right] + B_1' q^{1/2}(z)\exp\left[-i\int\limits_{(t_1')}^{z} q(z)dz\right]$$

$$(3.48.3)$$

with

$$\begin{pmatrix} A_1' \\ B_1' \end{pmatrix} = \begin{pmatrix} \exp(K) & i[\exp(2K)+1]^{1/2}\exp(+i\phi) \\ i[\exp(2K)+1]^{1/2}\exp(-i\phi) & -\exp(K) \end{pmatrix}\begin{pmatrix} A_1'' \\ 0 \end{pmatrix}$$

$$= \begin{pmatrix} \exp(K)A_1'' \\ i[\exp(2K)+1]^{1/2}\exp(-i\phi)A_1'' \end{pmatrix} \qquad (3.48.4)$$

and, according to (3.44.40) with the bar omitted,

$$K = -\frac{1}{2i} \int_{\Lambda} q(z)dz. \tag{3.48.5}$$

Here Λ is a closed contour, encircling t_1' and t_1'', along which the integration is performed in the positive direction. The wave function (3.48.3) can also be written as

$$\psi(z) = A_0'' q^{-1/2}(z) \exp\left[-i \int_{(t_0'')}^{z} q(z)dz \right] + B_0'' q^{-1/2}(z) \exp\left[+i \int_{(t_0'')}^{z} q(z)dz \right],$$
$$\tag{3.48.6}$$

where

$$\begin{pmatrix} A_0'' \\ B_0'' \end{pmatrix} = \begin{pmatrix} B_1'\exp(-iL) \\ A_1'\exp(+iL) \end{pmatrix} = \begin{pmatrix} 0 & \exp(-iL) \\ \exp(+iL) & 0 \end{pmatrix} \begin{pmatrix} A_1' \\ B_1' \end{pmatrix} \tag{3.48.7}$$

with

$$L = \int_{(t_1')}^{(t_0'')} q(z)dz, \quad \mathrm{Re}L > 0, \quad -\mathrm{Re}L \ll \mathrm{Im}L < 0. \tag{3.48.8}$$

The anti-Stokes line emerging from t_1', on which (3.48.3) and (3.48.6) are valid, does not exactly coincide with the anti-Stokes lines emerging from t_0'' and proceeding close to the real axis, but these two anti-Stokes lines lie very close to each other. Therefore expression (3.48.6) for the wave function is also valid on the anti-Stokes line that emerges from t_0'' and lies close to the real axis (which means that it coincides with part of the real axis when E is real). Inserting (3.48.4) into (3.48.7), we get

$$\begin{pmatrix} A_0'' \\ B_0'' \end{pmatrix} = \begin{pmatrix} [\exp(2K) + 1]^{1/2} \exp[-i(L + \phi - \pi/2)]A_1'' \\ \exp(K + iL)A_1'' \end{pmatrix}, \tag{3.48.9}$$

and hence

$$A_0''/B_0'' = [1 + \exp(-2K)]^{1/2} \exp[-i(2L + \phi - \pi/2)]. \tag{3.48.10}$$

On the Stokes line that emerges from t_0'' and lies close to the real axis (which means that it coincides with part of the real axis when E is real), the wave function that is almost positive and vanishes at $z = 0$, is, except for an arbitrary constant factor (note the approximate value of the phase of $Q^{1/2}(z)$ to the left of t_0'' in

Fig. 3.48.1)

$$\psi(z) = \exp(3i\pi/4)q^{-1/2}(z)\exp\left[+i\int\limits_{(t_0'')}^{z}q(z)dz\right], \quad 0 < \mathrm{Re}\,z < \mathrm{Re}\,t_0'' \approx t_0''.$$

(3.48.11)

On the anti-Stokes line that emerges from t_0'' and lies close to the real axis, the same wave function is according to the connection formula (3.8.6) (with due regard to the difference between the phase of $q^{-1/2}(z)$ in Section 3.8 and in the present problem)

$$\psi(z) = \exp(3i\pi/4) \times 2\exp(-i\pi/4)q^{-1/2}(z)\cos\left[\int\limits_{(t_0'')}^{z}-q(z)dz - \pi/4\right]$$

$$= 2iq^{-1/2}(z)\cos\left[\int\limits_{(t_0'')}^{z}q(z)dz + \pi/4\right]$$

$$= \exp(i\pi/4)q^{-1/2}(z)\exp\left[-i\int\limits_{(t_0'')}^{z}q(z)dz\right]$$

$$+ \exp(3i\pi/4)q^{-1/2}(z)\exp\left[+i\int\limits_{(t_0'')}^{z}q(z)dz\right]. \qquad (3.48.12)$$

Identifying (3.48.6) and (3.48.12), we obtain

$$A_0'' = \exp(i\pi/4), \qquad (3.48.13a)$$
$$B_0'' = \exp(3i\pi/4), \qquad (3.48.13b)$$

and hence

$$A_0''/B_0'' = -i. \qquad (3.48.14)$$

Inserting (3.48.14) into (3.48.10), we get

$$\exp[2i(L + \phi/2 - \pi/2)] = [1 + \exp(-2K)]^{1/2}, \qquad (3.48.15)$$

i.e.,

$$L = (s + 1/2)\pi - \tfrac{1}{2}\phi - \tfrac{1}{4}i\,\ln[1 + \exp(-2K)]$$
$$\approx (s + 1/2)\pi - \tfrac{1}{2}\phi - \tfrac{1}{4}i\,\exp(-2K), \qquad (3.48.16)$$

where s is an integer, which must be non-negative since $\mathrm{Re}\,L > 0$, $K \gg 1$ and

$|\phi| \ll 1$. From (3.48.16) we obtain the approximate formula

$$(L)_{E=\mathrm{Re}E_s} + \left(\frac{dL}{dE}\right)_{E=\mathrm{Re}E_s} (E_s - \mathrm{Re}E_s) = (s+1/2)\pi - \tfrac{1}{2}\phi - \tfrac{1}{4}i\exp(-2K).$$

(3.48.17)

Noting that $E_s - \mathrm{Re}E_s = i\,\mathrm{Im}E_s$ and separating (3.48.17) into real and imaginary parts, we obtain

$$(L)_{E=\mathrm{Re}E_s} = (s+1/2)\pi - \tfrac{1}{2}\mathrm{Re}\phi,$$

(3.48.18a)

$$\left(\frac{dL}{dE}\right)_{E=\mathrm{Re}E_s} \mathrm{Im}E_s = -\tfrac{1}{2}\mathrm{Im}\phi - \tfrac{1}{4}\exp(-2K).$$

(3.48.18b)

Using (3.44.33), (2.5.13b) with $\bar{K}_0$ replaced by $\bar{K}$, and (3.44.32b), we obtain for ϕ in the first-order approximation the expression

$$\phi \approx -\phi^{(3)} = \frac{\pi}{24K}$$

from which we obtain approximately

$$\phi_{E=E_s} - \phi_{E=\mathrm{Re}E_s} = -\left(\frac{\pi}{24K^2}\frac{dK}{dE}\right)_{E=\mathrm{Re}E_s} (E_s - \mathrm{Re}E_s)$$

$$= -\left(\frac{\pi}{24K^2}\frac{dK}{dE}\right)_{E=\mathrm{Re}E_s} i\,\mathrm{Im}E_s,$$

and hence, since ϕ is real for $E = \mathrm{Re}E_s$,

$$\mathrm{Im}\phi \approx -\left(\frac{\pi}{24K^2}\frac{dK}{dE}\right)_{E=\mathrm{Re}E_s} \mathrm{Im}E_s, \quad \text{first-order approximation.}$$

(3.48.19)

Inserting (3.48.19) into (3.48.18b), we obtain approximately

$$\mathrm{Im}E_s = -\Gamma/2,$$

(3.48.20a)

$$\Gamma = \frac{\exp(-2K)}{2\dfrac{dL}{dE} - \dfrac{\pi}{24K^2}\dfrac{dK}{dE}} = \frac{\exp(-2K)}{2\dfrac{d}{dE}\left(L + \dfrac{\pi}{48K}\right)},$$

(3.48.20b)

where the quantities on the right-hand side of (3.48.20b) are to be evaluated for $E = \mathrm{Re}E_s$. The denominator in (3.48.20b) is positive since $dL/dE > 0$ and $dK/dE < 0$ for $E = E_s$. The second term in the denominator of (3.48.20b) is due to the quantity ϕ, which thus causes a slight decrease of the value of Γ.

When one treats the Stark effect of a hydrogenic atom or ion, one arrives at two coupled differential equations of the Schrödinger type, in one of which a potential barrier appears in a similar way to that in the present problem. It is therefore of interest to say here something about the great accuracy that can sometimes be achieved in applications of the phase-integral method to the Stark effect of a hydrogen atom. For a state of a hydrogen atom with the principal quantum number 25 and the magnetic quantum number unity in an electric field of 2514 V/cm Fröman and Fröman (1984) obtained by means of the phase-integral method the energy to six digits in just the third-order approximation. For the same state Silverstone and Koch (1979) had obtained the energy to six digits using 24th-order Rayleigh–Schrödinger perturbation theory combined with a [12/12] Padé approximant, but the phase-integral results indicated that the last one of these digits was wrong by one unit. For a state of a hydrogen atom with the principal quantum number 30 and the magnetic quantum number zero in an electric field of 800 V/cm Fröman and Fröman (1984) obtained by means of the phase-integral method the energy to nine digits in just the third-order approximation. For the same state Damburg and Kolosov (private communication to N. Fröman in a letter of 22 February 1985) obtained by means of an efficient, purely numerical method the same energy to nine digits, but the phase-integral results indicated that the last one of these digits was wrong by one unit. For the same state Silverstone and Koch (1979) had earlier obtained the energy to six digits using 24th-order Rayleigh–Schrödinger perturbation theory combined with a [12/12] Padé approximant, but the last digit was wrong by three units.

3.49 Transmission coefficient for a particle penetrating a real single-hump potential barrier

Use the connection formula for a real single-hump potential barrier to derive a formula for the transmission coefficient for a non-relativistic quantal particle penetrating such a barrier; see Fig. 3.45.1. Apply the resulting formula to calculate in the first-order approximation the transmission coefficient for a quantal particle of energy E penetrating the parabolic potential barrier

$$V(x) = V_0(1 - x^2/a^2), \quad V_0 > 0. \tag{3.49.1}$$

Solution. In the connection formula for a real single-hump potential barrier, given by (3.45.5a,b), (3.45.15) and (3.45.9a), we put $B'' = 0$ and get

$$\psi(x'') = \tilde{A}'' |q^{-1/2}(x'')| \exp\left[+i \left| \mathrm{Re} \int_{(t'')}^{x''} q(z)dz \right| \right], \tag{3.49.2}$$

which represents the transmitted wave (moving away from the barrier), and

$$\psi(x') = \tilde{A}'|q^{-1/2}(x')|\exp\left[+i\left|\operatorname{Re}\int_{(t')}^{x'} q(z)dz\right|\right]$$

$$+ \tilde{B}'|q^{-1/2}(x')|\exp\left[-i\left|\operatorname{Re}\int_{(t')}^{x'} q(z)dz\right|\right], \qquad (3.49.3)$$

where the second term represents the incident wave (moving towards the barrier), and the first term represents the reflected wave (moving away from the barrier), and where according to (3.45.15) and (3.45.9a)

$$\begin{pmatrix} \tilde{A}' \\ \tilde{B}' \end{pmatrix} = \begin{pmatrix} -i\exp(K) & [\exp(2K)+1]^{1/2}\exp(+i\tilde{\phi}) \\ [\exp(2K)+1]^{1/2}\exp(-i\tilde{\phi}) & i\exp(K) \end{pmatrix}\begin{pmatrix} \tilde{A}'' \\ 0 \end{pmatrix}$$

$$= \begin{pmatrix} -i\exp(K)\tilde{A}'' \\ [\exp(2K)+1]^{1/2}\exp(-i\tilde{\phi})\tilde{A}'' \end{pmatrix}, \qquad (3.49.4)$$

with K defined according to (3.45.11), i.e.,

$$K = \frac{1}{2i}\int_{\Lambda} q(z)dz, \qquad (3.49.5)$$

where Λ is a closed contour encircling t' and t'', along which the integration is performed in the positive direction. From (3.49.4) we obtain

$$\tilde{A}''/\tilde{B}' = [\exp(2K)+1]^{-1/2}\exp(+i\tilde{\phi}). \qquad (3.49.6)$$

In a classically allowed region the classical velocity of the particle is $\hbar|R^{1/2}(x)|/m$, and, with a reasonable choice of the base function $Q(x)$ this approaches $\hbar|q(x)|/m$ as $x \to \pm\infty$, where the higher-order terms in $q(x)$ are negligible. For the transmitted wave the probability density is $|\tilde{A}''^2/q(x'')|$, and the probability current is $[\hbar|q(x'')|/m]|\tilde{A}''^2/q(x'')| = \hbar|\tilde{A}^2|/m$. Similarly the probability current for the incident wave is $\hbar|\tilde{B}'|^2/m$. The transmission coefficient T is therefore

$$T = \frac{\hbar|\tilde{A}''|^2/m}{\hbar|\tilde{B}'|^2/m} = \left|\frac{\tilde{A}''}{\tilde{B}'}\right|^2 = \frac{1}{\exp(2K)+1}, \qquad (3.49.7)$$

where the last equality has been obtained by means of (3.49.6). It should be noted that by using a connection formula at each one of the simple turning points instead of the connection formula for a barrier, one arrives at the less accurate formula $T = \exp(-2K)$, which one obtains by neglecting unity compared with $\exp(2K)$ in the denominator of (3.49.7).

Table 3.49.1. *The transmission coefficient T as a function of energy for a real, symmetric Eckart–Epstein barrier. The value $E/V_{\max} = 1.0$ corresponds to the top of the barrier. (From N. Fröman (1980).)*

$E/V_{\max}$	First-order approximation	Third-order approximation	Exact value
0.4	0.240	0.603	0.576
0.6	0.330	0.7032	0.7034
0.8	0.418	0.775	0.786
1.0	0.500	0.828	0.841
1.2	0.574	0.867	0.880
1.4	0.640	0.895	0.907

For a real potential barrier that need not be well isolated, which means that the two turning points t' and t'' need not lie far away from other existing transition points, we obtain instead of (3.49.7) the formula

$$T = 1/(\theta^2 + 1), \tag{3.49.8}$$

where θ, defined in (2.5.1a), is only roughly equal to $\exp(K)$; see (2.5.10a). When the transition points that are not associated with the barrier move away from the barrier, θ approaches $\exp(K)$.

In the first-order approximation we have $K > 0$ for energies below the top of the barrier and $K < 0$ for energies above the top of the barrier. Hence, in the first-order approximation $T < 1/2$ for subbarrier energies. However, if higher-order approximations are used, K may also be negative for energies below the top of the barrier, and it turns out that the transmission cofficient may also be obtained with the aid of higher-order approximations for thin barriers for which the first-order approximation is not applicable at all. An example is displayed in Table 3.49.1.

Parabolic barrier. For a parabolic barrier the only zeros of $Q^2(z)$ are the two turning points t' and t'', and we expect the formula (3.49.7) to be exact, which is readily verified as follows. Using (3.49.1) we obtain from (3.49.5), in the first-order approximation and with the choice $Q^2(x) = R(x)$,

$$K = \frac{1}{2i} \int_\Lambda \left\{ \frac{2m}{\hbar^2} [E - V(z)] \right\}^{1/2} dz = \left(\frac{m}{2\hbar^2} \right)^{1/2} \int_\Lambda [V_0(1 - z^2/a^2) - E]^{1/2} dz$$

$$= \left[\frac{m}{2\hbar^2} (V_0 - E) \right]^{1/2} \int_\Lambda (1 - z^2/t^2)^{1/2} dz, \tag{3.49.9}$$

where

$$t = -t' = t'' = a[(V_0 - E)/V_0]^{1/2}. \tag{3.49.10}$$

Evaluating the integral in (3.49.9) by residue calculus, we obtain with due regard
to (3.49.10)

$$K = \left[\frac{2m}{\hbar^2}(V_0 - E)\right]^{1/2} \pi t/2 = \frac{\pi a\,(V_0 - E)}{\hbar}\left(\frac{m}{2V_0}\right)^{1/2}. \quad (3.49.11)$$

The result for the transmission coefficient T obtained from (3.49.7) and (3.49.11)
agrees with the exact value for the parabolic barrier. The higher-order approximations give no contributions to T.

3.50 Transmission coefficient for a particle penetrating a real, *symmetric*, superdense double-hump potential barrier

Find the transmission coefficient for a quantal particle penetrating a real, symmetric,
superdense double-hump potential barrier. The generalization to the situation in
which the barriers may be either superdense or underdense is straightforward and
is therefore not considered in this section; see formulas in Section 3.45.

Solution. The transition zeros, which lie on the real axis since we are considering
subbarrier penetration, are $t_0'(= -t_1'')$, $t_0''(= -t_1')$, t_1' and $t_1''(> t_1')$; see Fig. 3.50.1.
We assume that t_0'' and t_1' are well separated, while t_0' and t_0'' (and hence t_1' and t_1'')
may lie at an arbitrary distance from each other.

We shall calculate the transmission coefficient for a quantal particle of real energy.
For $x > t_1''$ the wave function representing a transmitted wave (moving to the right)
is (cf. (3.49.2))

$$\psi(x) = \tilde{A}_1''|q^{-1/2}(x)|\exp\left[+i\left|\int_{(t_1'')}^{x} q(z)dz\right|\right], \quad x > t_1''. \quad (3.50.1)$$

For $t_0'' < x < t_1'$ we then have (cf. (3.49.3))

$$\psi(x) = \tilde{A}_1'|q^{-1/2}(x)|\exp\left[+i\left|\int_{(t_1')}^{x} q(z)dz\right|\right]$$
$$+ \tilde{B}_1'|q^{-1/2}(x)|\exp\left[-i\left|\int_{(t_1')}^{x} q(z)dz\right|\right], \quad t_0'' < x < t_1', \quad (3.50.2)$$

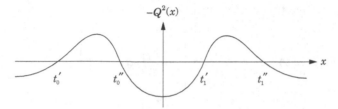

Figure 3.50.1 Symmetric double-hump potential barrier.

where according to (3.49.4), with slightly changed notation,

$$\begin{pmatrix} \tilde{A}'_1 \\ \tilde{B}'_1 \end{pmatrix} = \begin{pmatrix} \exp(K - i\pi/2)\tilde{A}''_1 \\ [\exp(2K) + 1]^{1/2} \exp(-i\tilde{\phi})\tilde{A}''_1 \end{pmatrix} \qquad (3.50.3)$$

with (cf. (3.49.5))

$$K = \left| \int_{(t'_1)}^{(t''_1)} q(z)dz \right| = \left| \int_{(t'_0)}^{(t''_0)} q(z)dz \right|, \qquad (3.50.4)$$

the last equality being needed in the derivation of (3.50.9) and (3.50.10) below. The wave function (3.50.2) can be written as

$$\psi(x) = \tilde{A}''_0 |q^{-1/2}(x)| \exp\left[+i \left| \int_{(t''_0)}^{x} q(z)dz \right| \right]$$

$$+ \tilde{B}''_0 |q^{-1/2}(x)| \exp\left[-i \left| \int_{(t''_0)}^{x} q(z)dz \right| \right], \quad t''_0 < x < t'_1, \quad (3.50.5)$$

where

$$\tilde{A}''_0 = \tilde{B}'_1 \exp(-iL), \qquad (3.50.6a)$$
$$\tilde{B}''_0 = \tilde{A}'_1 \exp(+iL), \qquad (3.50.6b)$$

with

$$L = \left| \int_{(t''_0)}^{(t'_1)} q(z)dz \right|. \qquad (3.50.7)$$

Inserting the expressions for $\tilde{A}'_1$ and $\tilde{B}'_1$ according to (3.50.3) into (3.50.6a,b), we obtain

$$\tilde{A}''_0 = [\exp(2K) + 1]^{1/2} \exp[-i(L + \tilde{\phi})]\tilde{A}''_1, \qquad (3.50.8a)$$
$$\tilde{B}''_0 = \exp[K + i(L - \pi/2)]\tilde{A}''_1. \qquad (3.50.8b)$$

For $x < t_0'$ the wave function is given by the formula

$$\psi(x) = \tilde{A}_0' |q^{-1/2}(x)| \exp\left[+i \left| \int_{(t_0')}^x q(z)dz \right| \right]$$

$$+ \tilde{B}_0' |q^{-1/2}(x)| \exp\left[-i \left| \int_{(t_0')}^x q(z)dz \right| \right], \quad x < t_0', \quad (3.50.9)$$

where according to the connection formula for a real potential barrier, contained in (3.45.5a,b), (3.45.15) and (3.45.9a),

$$\begin{pmatrix} \tilde{A}_0' \\ \tilde{B}_0' \end{pmatrix} = \begin{pmatrix} \exp(K - i\pi/2) & [\exp(2K) + 1]^{1/2} \exp(+i\tilde{\phi}) \\ [\exp(2K) + 1]^{1/2} \exp(-i\tilde{\phi}) & \exp(K + i\pi/2) \end{pmatrix} \begin{pmatrix} \tilde{A}_0'' \\ \tilde{B}_0'' \end{pmatrix}.$$

$$(3.50.10)$$

Inserting (3.50.8a,b) into (3.50.10), we obtain

$$\tilde{A}_0'/\tilde{A}_1'' = 2i \exp(K)[\exp(2K) + 1]^{1/2} \sin(L + \tilde{\phi} - \pi/2), \quad (3.50.11a)$$

$$\begin{aligned} \tilde{B}_0'/\tilde{A}_1'' &= \exp[-i(\pi/2 + \tilde{\phi})]\{[\exp(2K) + 1]\exp[+i(\pi/2 - L - \tilde{\phi})] \\ &\quad - \exp(2K)\exp[-i(\pi/2 - L - \tilde{\phi})]\} \\ &= \exp[-i(\pi/2 + \tilde{\phi})]\{\cos[\pi/2 - (L + \tilde{\phi})] \\ &\quad + i[2\exp(2K) + 1]\sin[\pi/2 - (L + \tilde{\phi})]\} \\ &= \{1 + 4\exp(2K)[\exp(2K) + 1]\sin^2[\pi/2 - (L + \tilde{\phi})]\}^{1/2} \\ &\quad \times \exp[i(\arctan\{[2\exp(2K) + 1]\tan[\pi/2 - (L + \tilde{\phi})]\} - \pi/2 - \tilde{\phi})], \end{aligned}$$

where arctan lies in the same quadrant modulo 2π as $\pi/2 - (L + \tilde{\phi})$. Thus

$$\tilde{B}'_0/\tilde{A}_1'' = \hat{A}(E)\exp[i(\chi - \pi/2 - \tilde{\phi})], \quad (3.50.11b)$$

with

$$\hat{A}(E) = \{1 + 4\exp(2K)[\exp(2K) + 1]\sin^2[\pi/2 - (L + \tilde{\phi})]\}^{1/2}, \quad (3.50.12a)$$
$$\chi = \arctan\{[2\exp(2K) + 1]\tan[\pi/2 - (L + \tilde{\phi})]\}. \quad (3.50.12b)$$

From (3.50.11a,b) and (3.50.12a) we obtain

$$|\tilde{A}_0'/\tilde{A}_1''|^2 = 4\exp(2K)[\exp(2K) + 1]\sin^2(L + \tilde{\phi} - \pi/2), \quad (3.50.13a)$$
$$|\tilde{B}_0'/\tilde{A}_1''|^2 = 1 + 4\exp(2K)[\exp(2K) + 1]\sin^2(L + \tilde{\phi} - \pi/2). \quad (3.50.13b)$$

The transmission coefficient T is

$$T = \left| \frac{\tilde{A}''_1}{\tilde{B}'_0} \right|^2 = \frac{1}{|\tilde{B}'_0/\tilde{A}''_1|^2} = \frac{1}{1 + 4 \exp(2K)[\exp(2K) + 1] \sin^2(L + \tilde{\phi} - \pi/2)}$$

(3.50.14a)

in agreement with eqs. (40) and (41) in N. Fröman and Dammert (1970), and the reflection coefficient R is

$$R = \left| \frac{\tilde{A}'_0}{\tilde{B}'_0} \right|^2 = \frac{|\tilde{A}'_0/\tilde{A}''_1|^2}{|\tilde{B}'_0/\tilde{A}''_1|^2} = \frac{4 \exp(2K)[\exp(2K) + 1] \sin^2(L + \tilde{\phi} - \pi/2)}{1 + 4 \exp(2K)[\exp(2K) + 1] \sin^2(L + \tilde{\phi} - \pi/2)}.$$

(3.50.14b)

The transmission coefficient (3.50.14a) attains its maximum value, unity, when $L = (s + 1/2)\pi - \tilde{\phi}$, and it attains its minimum value, $[2 \exp(2K) + 1]^{-2}$, when $L = s\pi - \tilde{\phi}$, where s is an integer.

When the base function is chosen to be

$$Q(z) = \{2m[E - V(z)]/\hbar^2\}^{1/2},$$

(3.50.15)

and the first-order approximation is used, we have according to (3.50.7)

$$L = \int_{t''_0}^{t'_1} \{2m[E - V(z)]/\hbar^2\}^{1/2} dz.$$

(3.50.16)

Putting

$$E = E_s + (E - E_s),$$

(3.50.17)

where E_s is an energy for which the transmission coefficient attains its maximum value, unity, and assuming that $|E - E_s|$ is not too large, we obtain from (3.50.16) (see the calculations in Section 3.26)

$$L \approx \int_{t''_0}^{t'_1} \{2m[E_s - V(z)]/\hbar^2\}^{1/2} dz$$

$$+ \left(\frac{d}{dE} \int_{t''_0}^{t'_1} \{2m[E - V(z)]/\hbar^2\}^{1/2} dz \right)_{E=E_s} (E - E_s)$$

$$= (s + 1/2)\pi - \tilde{\phi} + \frac{1}{\hbar} \int_{t''_0}^{t'_1} \frac{dz}{\{2[E_s - V(z)]/m\}^{1/2}}(E - E_s), \quad (3.50.18)$$

i.e.,

$$L = (s + 1/2)\pi - \tilde{\phi} + \frac{E - E_s}{2\hbar}\tau = (s + 1/2)\pi - \tilde{\phi} + \frac{\pi(E - E_s)}{h\nu},$$

$$(3.50.19)$$

where

$$\tau = 2 \int_{t_0''}^{t_1'} \frac{dz}{\{2[E_s - V(z)]/m\}^{1/2}}$$

$$(3.50.20)$$

is the time for a full oscillation of a classical particle with the energy E_s in the potential well, and $\nu = 1/\tau$ is the corresponding frequency. Inserting (3.50.19) into (3.50.14a), we obtain

$$T = \frac{1}{1 + 4\exp(2K)[\exp(2K) + 1]\sin^2 \dfrac{\pi(E - E_s)}{h\nu}}.$$

$$(3.50.21)$$

When $E - E_s$ is an integer multiple of $h\nu/\pi$, the double-hump potential is completely transparent ($T = 1$), but when $E - E_s - \pi/2$ is an integer multiple of $h\nu/\pi$, the transmission coefficient has minima equal to $1/[2\exp(2K) + 1]^2 \approx (1/4)\exp(-2K)$. The transmission coefficient for a thick single-hump potential barrier is approximately $\exp(-2K)$ according to Section 3.49.

References

Abawi, A. T., Dashen, R. F., and Levine, H., 1997, *J Math Phys* **38**, 1623–1649.

Abramowitz, M., and Stegun, I. A., Editors, 1965, *Handbook of Mathematical Functions*. Applied Mathematics Series **55**. Fourth Printing 1965, with corrections, National Bureau of Standards, Washington DC.

de Alfaro, V., and Regge, T., 1965, *Potential Scattering*. North-Holland, Amsterdam.

Amaha, A., 1993, *Molec Phys* **78**, 345–356.

Bateman Manuscript Project, 1953, *Higher Transcendental Functions*, Vols I and II, McGraw-Hill Book Company, New York.

Bateman Manuscript Project, 1954, *Tables of Integral Transforms*, Vols I and II, McGraw-Hill Book Company, New York.

Bateman Manuscript Project, 1955, *Higher Transcendental Functions*, Vol III, McGraw-Hill Book Company, New York.

Berry, M. V., and Mount, K. E., 1972, *Rep Prog Phys* **35**, 315–397.

Bertocchi, L., Fubini, S., and Furlan, G., 1965, *Nuovo Cim* **35**, 599–632.

Birkhoff, G. D., 1908, *Trans Am Math Soc* **9**, 219–231.

Birkhoff, G. D., 1933, *Bull Am Math Soc* **39**, 681–700.

Blumenthal, O., 1912, *Archiv Math Phys (Dritte Reihe)* **19**, 136–174.

Brillouin, L., 1926a, *C R Acad Sci (Paris)* **183**, 24–26.

Brillouin, L., 1926b, *J Phys Radium (Série VI)* **7**, 353–368.

Broer, L. J. F., 1963, *Appl Sci Res* **B10**, 110–118.

Campbell, J. A., 1972, *J Comp Phys* **10**, 308–315.

Campbell, J. A., 1979, *J Phys A: Math Gen* **12**, 1149–1154.

Carlini, F., 1817, *Ricerche sulla convergenza della serie che serve alla soluzione del problema di Keplero*. Appendice all' Effemeridi Astronomiche di Milano per l'Anno 1818, pp. 3–48. (Review: *Giornale di Fisica, Chimica, Storia Naturale Medicina ed Arti*. X, 458–460, 1817).

Carlini, F., 1850, *Astron Nachr* **30**, 197–254. Translation into German and revision by Jacobi of the paper by Carlini (1817); also published in *C. G. J. Jacobi's Gesammelte Werke*, edited by K. Weierstrass, Band **7**, pp 189–245, Berlin 1891.

Cherry, T. M., 1950, *Trans Am Math Soc* **68**, 224–257.

Child, M. S., 1974, *Molecular Collision Theory*. Academic Press, London. Reprinted with corrections 1984.

Choi, S., and Ross, J., 1962, *Proc Nat Acad Sci* **48**, 803–806.

Cole, M. W., and Good, Jr, R. H., 1978, *Phys Rev* **A18**, 1085–1088.

Connor, J. N. L., 1968, *Mol Phys* **15**, 37–46.

Copson, E. T., 1965, *Asymptotic Expansions*. Cambridge University Press, Cambridge.

Dagens, L., 1969, *J Physique (Paris)* **30**, 593–597.

Dammert, Ö., and Fröman, P. O., 1980, *J Math Phys* **21**, 1683–1687.

Debye, P., 1909, *Math Ann* **67**, 535–558.

Delves, L. M., 1963, *Nucl Phys* **41**, 497–503.

Dingle, R. B., 1965, *Proc Phys Soc* **86**, 1366–1368.

Dingle, R. B., 1973, *Asymptotic Expansions: Their Derivation and Interpretation*. Academic Press, London.

Erdélyi, A., 1956, *Proc Int Congr Math 1954*, **III**, 92–101.

Erdélyi, A., 1960, *J Math Phys* **1**, 16–26.

Flügge, S., 1974, *Practical Quantum Mechanics*, Vol. I. Springer-Verlag, New York.

Ford, K. W., Hill, D. L., Wakano, M., and Wheeler, J. A., 1959, *Ann Phys (NY)* **7**, 239–258.

Fowler, R. H., Gallop, E. G., Lock, C. N. H., and Richmond, W. H., 1921, *Phil Trans Roy Soc London* **A221**, 295–387.

Fröman, N., 1966a, *Ark Fys* **31**, 381–408.

Fröman, N., 1966b, *Ark Fys* **31**, 445–451.

Fröman, N., 1966c, *Ark Fys* **32**, 79–97.

Fröman, N., 1966d, *Ark Fys* **32**, 541–548.

Fröman, N., 1970, *Ann Phys (NY)* **61**, 451–464.

Fröman, N., 1974, *Phys Lett* **48A**, 137–139.

Fröman, N., 1978a, *Phys Rev* **A17**, 493–504.

Fröman, N., 1978b, *J Math Phys* **19**, 1141–1146.

Fröman, N., 1980, *Semiclassical and Higher-Order Approximations: Properties. Solution of Connection Problems*. Proceedings of the NATO Advanced Study Institute, Cambridge, England, in September 1979, on Semiclassical Methods in Molecular Scattering and Spectroscopy, pp. 1–44, edited by M. S. Child. NATO Advanced Study Institute Series C-Mathematical and Physical Sciences, Volume 53. Reidel, Dordrecht.

Fröman, N., and Dammert, Ö., 1970, *Nucl Phys* **A147**, 627–649.

Fröman, N., and Fröman, P.O., 1965, *JWKB Approximation, Contributions to the Theory*. North-Holland, Amsterdam. Russian translation: MIR, Moscow 1967.

Fröman, N., and Fröman, P. O., 1970, *Nucl Phys* **A147**, 606–626.

Fröman, N., and Fröman, P. O., 1972, *Phys Rev* **A6**, 2064–2067.

Fröman, N., and Fröman, P. O., 1974a, *Ann Phys (NY)* **83**, 103–107. (Review: *Zentralblatt für Mathematik und ihre Grenzgebiete*, Mathematics Abstracts **279**, 190–191, 1974.)

Fröman, N., and Fröman, P. O., 1974b, *Nuovo Cim* **20B**, 121–132.

Fröman, N., and Fröman, P. O., 1977, *J Math Phys* **18**, 903–906.

Fröman, N., and Fröman, P. O., 1978, *J Math Phys* **19**, 1823–1829.

Fröman, N., and Fröman, P. O., 1981, *J Physique (Paris)* **42**, 1491–1504.

Fröman, N., and Fröman, P. O., 1984, *Phase-Integral Calculation with Very High Accuracy of the Stark Effect in a Hydrogen Atom*. Colloque du 10 à 15 septembre 1984, CIRM (Luminy), sur *Méthodes Semi-Classiques en Mécanique Quantique*, p. 45, edited by D. Helffin and D. Robert. Publications de l'Université de Nantes, Institut de Mathématiques et d'Informatiques, 2 rue de la Houssinière, 44072 Nantes CEDEX, France.

Fröman, N., and Fröman, P. O., 1985a, *On the History of the So-called WKB-Method from 1817 to 1926*. Proceedings of the Niels Bohr Centennial Conference, Copenhagen 25–28 March 1985 on Semiclassical Descriptions of Atomic and Nuclear Collisions, pp 1–7, edited by J. Bang and J. de Boer. North-Holland, Amsterdam.

Fröman, N., and Fröman, P. O., 1985b, *Ann Phys (NY)* **163**, 215–226.

Fröman, N., and Fröman, P. O., 1989, *Int J Quant Chem* **35**, 751–760.

Fröman, N., and Fröman, P. O., 1996, *Phase-Integral Method Allowing Nearlying Transition Points*, With adjoined papers by A. Dzieciol, N. Fröman, P. O. Fröman,

A. Hökback, S. Linnæus, B. Lundborg, and E. Walles. Springer Tracts in Natural Philosophy Vol 40, edited by C. Truesdell. Springer-Verlag, New York.

Fröman, N., and Fröman, P. O., 1998, *J Math Phys* **39**, 4417–4429.

Fröman, N., and Myhrman, U., 1970, *Ark Fys* **40**, 497–508.

Fröman, N., Fröman, P. O., and Karlsson, F., 1979, *Molec Phys* **38**, 749–767.

Fröman, N., Fröman, P. O., and Larsson, K., 1994, *Phil Trans R Soc Lond* **A347**, 1–22.

Fröman, N., Fröman, P. O., and Lundborg, B., 1988a, *Math Proc Camb Phil Soc* **104**, 153–179.

Fröman, N., Fröman, P. O., and Lundborg, B., 1988b, *Math Proc Camb Phil Soc* **104**, 181–191.

Fröman, N., Fröman, P. O., and Lundborg, B., 1996, adjoined paper (Chapter 5) in Fröman and Fröman (1996).

Fröman, N., Fröman, P. O., Andersson, N., and Hökback, A., 1992, *Phys Rev* **D45**, 2609–2616.

Fröman, N., Fröman, P. O., Myhrman, U., and Paulsson, R., 1972, *Ann Phys (NY)* **74**, 314–323.

Fröman, P. O., 1957, *Mat Fys Skr Dan Vid Selsk* **1**, No. 3.

Fröman, P. O., 1974, *Ann Phys (NY)* **88**, 621–630.

Fröman, P. O., 2000, *J Math Phys* **41**, 7952–7963.

Fröman, P. O., Karlsson, F., and Yngve, S., 1986, *J Math Phys* **27**, 2738–2747.

Fröman, P. O., Yngve, S., and Fröman, N., 1987, *J Math Phys* **28**, 1813–1826.

Fröman, P. O., Hökback, A., Walles, E., and Yngve, S., 1985, *Ann Phys (NY)* **163**, 252–264.

Fulling, S. A., 1983, *SIAM J Math Anal* **14**, 780–795.

Furry, W. H., 1947, *Phys Rev* **71**, 360–371.

Gans, R., 1915, *Ann Physik (Vierte Folge)* **47**, 709–736.

Glaser, W., and Braun, G., 1954, *Act Phys Austriaca* **9**, 41–74.

Glaser, W., and Braun, G., 1955, *Act Phys Austriaca* **9**, 267–296.

Green, G., 1837, *Trans Camb Phil Soc* **6**, 457–462.

Gustafson, S.-Å., and Lindahl, S., 1977, *J Comp Phys* **24**, 81–95.

ter Haar, D., Editor, 1964, *Selected Problems in Quantum Mechanics*. Revised and augmented second edition of Gol'dman, Krivchenkov, Kogan and Galitskii, *Problems in Quantum Mechanics*. Academic Press, New York.

Heading, J., 1962, *An Introduction to Phase-Integral Methods*. Methuen's Monographs on Physical Subjects, London and New York. Russian translation with an appendix by V. P. Maslov concerning the WKB method in the multi-dimensional case: MIR, Moscow 1965.

Horn, J., 1899a, *Math Ann* **52**, 271–292.

Horn, J., 1899b, *Math Ann* **52**, 340–362.

Jacobi, C. G. J., 1849, *Astron Nachr* **28**, 257–270; also published in C. G. J. Jacobi's *Gesammelte Werke*, edited by K. Weierstrass, Band **7**, pp 175–188, Berlin 1891.

Jeffreys, H., 1915, *Mem Roy Astr Soc* **60** (Part VI), 187–217.

Jeffreys, H., 1925, *Proc Lond Math Soc (Second Series)* **23**, 428–436.

Jeffreys, H., 1956, *Proc Camb Phil Soc* **52**, 61–66.

Karlsson , F., and Jedrzejek, C., 1987, *J Chem Phys* **86**, 3532–3538.

Kemble, E. C., 1935, *Phys Rev* **48**, 549–561.

Kemble, E. C., 1937, *The Fundamental Principles of Quantum Mechanics*. McGraw-Hill, New York. Reissue: Dover Publications, New York 1958.

Kramers, H. A., 1926, *Z Phys* **39**, 828–840.

Kraus, L., and Levine, L. M., 1961, *Comm Pure Appl Math* **14**, 49–68.

Krieger, J. B., 1969, *J Math Phys* **10**, 1455–1458.

Krieger, J. B., and Rosenzweig, C., 1967, *Phys Rev* **164**, 171–173.

Lamb, H., 1895, *Hydrodynamics*. Cambridge University Press, Cambridge.

Langer, R. E., 1934, *Bull Am Math Soc* **40**, 545–582.

Larsson, K., and Fröman, P. O., 1994, *Phil Trans R Soc Lond* **A347**, 23–35.

Linnæus, S., and Düring, M., 1985, *Ann Phys (NY)* **164**, 506–515.

Liouville, J., 1837, *J Math Pure Appl* **2**, 16–35.

Luke, Y. L., 1969, *The Special Functions and Their Approximations*, Vol. I. Academic Press, New York.

Lundborg, B., 1977, *Math Proc Camb Phil Soc* **81**, 463–483.

Maslov, V. P., and Fedoriuk, M. V., 1981, *Semi-Classical Approximation in Quantum Mechanics*. Reidel Publishing Company, Dordrecht.

McHugh, J. A. M., 1971, *Arch History Exact Sci* **7**, 277–324.

Messiah, A., 1959, *Mécanique Quantique*, Vol. 1. Dunod, Paris. English translation: *Quantum Mechanics*, Vol. 1. North-Holland, Amsterdam. Fourth printing 1967.

Olver, F. W. J., 1965a, *J Res Nat Bur Standards* **69B**, 271–290.

Olver, F. W. J., 1965b, *J Res Nat Bur Standards* **69B**, 291–300.

Paulsson, R., 1985, *Ann Phys (NY)* **163**, 245–251.

Paulsson, R., and Fröman, N., 1985, *Ann Phys (NY)* **163**, 227–244.

Rayleigh, Lord, (J. W. Strutt), 1912, *Proc R Soc London* **A86**, 207–226.

Rosenzweig, C., and Krieger, J. B., 1968, *J Math Phys* **9**, 849–860.

Scheibner, W., 1856a, *Astron J* **4**, 177–182.

Scheibner, W., 1856b, *Berichte der Kgl Sächs Ges d Wiss zu Leipzig, Math-Phys Classe* **8**, 40–64.

Scheibner, W., 1880a, *Math Ann* **17**, 531–544.

Scheibner, W., 1880b, *Math Ann* **17**, 545–560.

Schlesinger, L., 1906, *C R Acad Sci (Paris)* **142**, 1031–1033.

Schlesinger, L., 1907, *Math Ann* **63**, 277–300.

Schlissel, A., 1977, *Arch History Exact Sci* **16**, 307–378.

Siebert, E., and Krieger, J. B., 1970, *J Math Phys* **11**, 3111–3115.

Silverstone, H. J., 1985, *Phys Rev Lett* **55**, 2523–2526.

Silverstone, H. J., and Koch, P. M., 1979, *J Phys B: Atom Molec Phys* **12**, L537–L541.

Skorupski, A. A., 1980, *Rep Math Phys* **17**, 161–187.

Skorupski, A. A., 1988, *J Math Phys* **29**, 1814–1823.

Soop, M., 1965, *Ark Fys* **30**, 217–229.

de Sparre, M., 1898, *Atti della R Acc dei Lincei (Ser V). Rendiconti. Classe di sc fis, mat e nat* 7:2, 111–117.

Streszewski, M., and Jedrzejek, C., 1988, *Phys Rev* **A37**, 645–648.

Thidé, B., 1980, *J Math Phys* **21**, 1408–1415.

Wentzel, G., 1926, *Z Phys* **38**, 518–529.

Wheeler, J. A., 1976, *Studies in Mathematical Physics: Essays in Honor of Valentine Bargmann*, pp 351–422, edited by E. H. Lieb, B. Simon and A. S. Wightman. Princeton University Press, Princeton.

Yngve, S., 1972, *J Math Phys* **13**, 324–331.

Yngve, S., 1986, *Phys Rev* **A33**, 96–104.

Yngve, S., 1988, *J Math Phys* **29**, 931–936.

Yngve, S., and Linnæus, S., 1986, *J Phys A: Math Gen* **19**, 3017–3031.

Zwaan, A., 1929, *Intensitäten im Ca-Funkenspektrum*. Academisch Proefschrift. Joh. Enschedé en Zonen, Harlem. This doctoral dissertation was also published in: *Arch Néerlandaises Sci Exactes Naturelles (Série IIIA)* **12**, 1–76.

Author index

References to the list of references are not made, and names associated with concepts, such as Bohr radius, Bessel functions, Schrödinger equation, etc., are not mentioned. As regards N. Fröman and P. O. Fröman references are made only to Chapter 1.

Abawi, A. T., 115
Abramowitz, M., 157
de Alfaro, V., 10
Amaha, A., 46
Andersson, N., 31

Bateman, H., 8
Berry, M. V., 9
Bertocchi, L., 14
Birkhoff, G. D., 5, 6, 8
Blumenthal, O., 5
Braun, G., 11
Brillouin, L., 7, 8
Broer, L. J. F., 14

Campbell, J. A., 16, 17
Carlini, F., 1–5, 7
Cauchy, A. L., 4
Cherry, T. M., 11
Child, M. S., 48
Choi, S., 10
Cole, M. W., 147
Connor, J. N. L., 52
Copson, E. T., 93, 94

Dagens, L., 10
Damburg, R. J., 197
Dammert, Ö., 14, 15, 17, 203
Dashen, R. F., 115
Debye, P., 3
Delves, L. M., 10
Dingle, R. B., 9
Düring, M., 152, 166

Encke, J. F., 4
Erdélyi, A., 11

Fedoriuk, M. V., 11
Flügge, S., 139

Ford, K. W., 11, 52
Fowler, R. H., 6, 7
Fröman, N., in Chapter 1: 1, 9–11
Fröman, P. O., in Chapter 1: 1, 3, 9–11
Fubini, S., 14
Fulling, S. A., 15
Furlan, G., 14
Furry, W. H., 8, 10

Gallop, E. G., 6
Gans, R., 1, 6, 7
Glaser, W., 11
Good, Jr, R. H., 147
Green, G., 3–5, 7
Gustafson, S.-Å., 170

ter Haar, D., 139
Heading, J., 9, 11
Hill, D. L., 11, 52
Hökback, A., 31, 166, 171
Horn, J., 5–7

Jacobi, C. G. J., 4
Jedrzejek, C., 171
Jeffreys, H., 1, 6–10, 76, 81

Karlsson, F., 3, 10, 166, 170, 171
Kemble, E. C., 8, 9
Koch, P. M., 197
Kolosov, V. V., 197
Kramers, H. A., 1, 3, 7, 8
Kraus, L., 115
Krieger, J. B., 10, 142

Lagrange, J. L., 4
Lamb, H., 5, 7
Langer, R. E., 9
Larsson, K., 18, 140, 142
Levine, H., 115

Levine, L. M., 115
Lindahl, S., 170
Linnæus, S., 152, 166, 171
Liouville, J., 3, 4
Lock, C. N. H., 6
Luke, Y. L., 89, 159
Lundborg, B., 11, 20, 26, 28, 36, 47, 51, 52, 58, 180

Maslov, V. P., 11
Maxwell, J. C., 6
McHugh, J. A. M., 1
Messiah, A., 14
Mount, K. E., 9
Myhrman, U., 11, 48, 185

Olver, F. W. J., 8, 10

Paulsson, R., 11, 48, 166, 171, 185

Rayleigh, Lord (J. W. Strutt), 1, 5–7
Regge, T., 10
Richmond, W. H., 6
Rosenzweig, C., 10, 142
Ross, J., 10

Scheibner, W., 4, 5
Schlesinger, L., 5, 6
Schlissel, A., 1
Siebert, E., 10
Silverstone, H. J., 9, 100, 197
Skorupski, A. A., 16
Soop, M., 52
de Sparre, M., 5, 6
Stegun, I. A., 157
Streszewski, M., 171
Swirles Jeffreys, B., 8

Thidé, B., 152, 166
Truesdell, C., 8

Wakano, M., 11, 52
Walles, E., 166, 171
Wentzel, G., 7, 8
Wheeler, J. A., 11, 52, 147

Yngve, S., 3, 10, 35, 131, 133, 152,
       166, 171

Zwaan, A., 8, 9

Subject index

a-coefficients, 23
  associated with Airy functions, 94–6
  change of, 24, 30
  differential equation for, 23
  in expression for exact wave function, 23
  integral equation for, 24
Airy differential equation solution represented at a
    fixed point by a pure phase-integral function,
    98–102
α, *see* parameters α, β and γ
amplitude and phase of wave function,
  complex, 13
  relation between, 12–14
anti-Stokes lines, *see* Stokes and anti-Stokes lines

base function, 14
  choice of, 18–19, 60–2
  that makes quantization condition exact, 140–1
  that makes the phase-integral approximation valid
    close to the origin and the radial phase shift
    equal to zero for free particle, 60–2, 121, 123
  determination of, 15, 17–19, 60–2
  non-uniqueness of, 18
basic estimates of F-matrix elements, 26–7
β, *see* parameters α, β and γ
black-hole normal modes, 31
bookkeeping parameter, 15
Borel summation of JWKB expansions, 100

Carlini approximation, 8, 21
classical turning point, definition of in generalized
    sense, 35
classically allowed region in generalized sense, 35
classically forbidden region,
  in generalized sense, 35
  of infinite thickness, 187, 188
coefficient function $R(z)$, 47
comparison equation technique, 83–8
  Stokes constants obtained by means of, 11
  supplementary quantity ϕ for complex barrier
    obtained by means of, 51, 180
  supplementary quantity $\tilde{\phi}$ for real barrier obtained
    by means of, 47, 51–2

complex amplitude, 13
complex barrier, definition of, 171
complex energy, 192
complex phase, 13
compound symbol for tracing solutions in the
    complex plane, 29–31
  orientation of, 30–1
  traffic rules for, 30–1
compressed atom, *see* displacement of energy levels
connection formula,
  associated with
    complex barrier, 171–81
    real barrier, 49–53, 181–5
    transition zero in the complex plane, 72, 73
    turning point on the real axis, 10, 75–6, 80–1
  derived by comparison equation technique, 85, 88
  derived for the Airy differential equation, 102–4
  disregard of one-directional nature of
    yielding correct result, 136
    yielding erroneous result, 76–7, 81–2, 100,
      102–3, 104, 137, 185
  for tracing s-electron wave function away from the
    origin, where there is a strong Coulomb
    singularity, 159–60
  one-directional nature of, 9, 77–9, 81–2
    demonstrated for Airy differential equation,
      78–9, 102–4
connection problem,
  formulation of, 22
  importance of precise formulation of, 9, 22
  in condensed mathematical form, 22, 24
  rigorous method for mastering, 9, 22
contour integral, 19
  short-hand notation for, 20

Debye's asymptotic formula, derived by
  Carlini, 3
  modern phase-integral technique, 3
determination of potential from
  energy spectrum of particle in one-dimensional
    single-well potential, 142–4
  energy spectrum of particle in radial single-well
    potential, 144–50

determination of potential from (*cont.*)
  energy dependence of transmission coefficient of
    barrier, 147
  energies and widths of quasi-stationary states, 147
differential equation,
  auxiliary, 15, 163
  original, 15, 163
displacement of energy levels due to
  change of boundary condition (compressed atom),
    130–5
  omission of parameter β, 135, 191–2
distance between nearlying energy levels, 123–5, 127,
    132–3, 135, 192
double-well potential,
  general, quantization condition for, 189–90
  symmetric, quantization condition for, 190–1

ε, definition of, 23
ε_0, definition of, 15
expectation value formula, not involving wave
    function, 10
  exact formula, 164
  phase-integral formula, 165
    explanation of its unexpectedly high accuracy, 166
    for kinetic energy, 168
    some cases in which it is exact in first order, 166–7

F-matrix,
  connecting points on opposite sides of turning
    point, 35–8
  expressed in terms of parameters α, β and γ, 39
  corresponding to encircling of simple transition
    zero, 26
  definition of, 24–5
  dependence of on lower limit of integration in the
    phase integral, 33
  determinant of, 25
  differential equation for, 25
  elements,
    basic estimates of, 26–7
    explicit expressions for, 24–5
  exact solution expressed with use of, 23–4
  expressed in terms of two solutions of the
    differential equation, 34
  for simplest non-monotonic path, 28
  general procedure for use of, 27–8
  general relations satisfied by, 25
  given by convergent series, 24–5
  inversion formula for, 25
  method for mastering connection problems, 14, 22
  multiplication rule for, 25
  symmetry relations for, 25–6, 37
form of wave function, 12

γ, see parameters α, β and γ
global problem, 21

Hamilton–Jacobi equation, 7
Hellmann–Feynman formula, 164

Jeffreys' connection fomulas, 8

Kramers' modification of first-order WKB
    approximation, 3, 7, 8

local problem, 21

matrix element,
  phase-integral formulas for, not involving wave
    functions, 10, 170
  great accuracy of, 170–171
μ-integral, definition of, 27, 41, 42

normalization factor expressed in terms of frequency
    of oscillation, 155
normalization integral, formula for, not involving
    wave function, 10
  exact formula, 152
  phase-integral formula, 154
normalized wave function, 155

one-directional nature of connection formulas, see
    connection formula

Padé approximant, 197
parameterization of F-matrix connecting points on
    opposite sides of turning point, 39
  power and limitation of, 45–6
  procedure for use of, 46
parameters α, β and γ,
  associated with Airy differential equation, 97–8
  changes of, 41–2
  definition of, 39
  estimates of, 39
  limiting values of, 43
  relations between parameters associated with
    turning points of symmetric barrier, 114
  situations in which γ remains in final phase-integral
    formula, 76–7, 81–2, 110–11, 137
phase and amplitude of wave function, *see* amplitude
    and phase of wave function
phase of wave function,
  change of due to change of boundary condition,
    107, 108–13, 120–1
  on opposite sides of symmetric barrier, 108–13
phase integral $w(z)$, 12–13,
  multi-dimensional, 11
  replaced by contour integral, 19–20
    short-hand notation for, 20
  three-dimensional, 11
phase-integral approximation generated from
    unspecified base function, 10, 13–17
  advantages of compared to WKB approximation,
    13–15, 17–18, 20–1, 23, 142
phase-integral functions $f_1(z)$ and $f_2(z)$, 22
  associated with Airy differential equation, 93
  constant Wronskian of, 23
  differential equation for, 23
phase integrand $q(z)$, 10, 13
  asymptotic, 13
  exact, 13
  simple zeros of in neighbourhood of transition
    zero, 17

phase-shift formula, 121–3
Planck's constant not always useful as small
    parameter, 15
potential, auxiliary, 163
potential barrier,
  complex,
    definition of, 171
    connection formula for, 178, 181
  real,
    connection formula for, 49, 52–3, 183–5
    Eckart–Epstein, 199
    erroneous connection matrix for, 185
probability density at the origin for s-electron,
    163
properties of phase-integral approximation along
    anti-Stokes line, 66
  path of strict monotonicity, in particular Stokes
    line, 66–9

q-equation,
  auxiliary, 15
  original, 13, 15
quantization condition,
  for particle in
    general double-well potential, 189–90
    potential with very steep walls, 140–2
    single-well potential, 115–17, 126–7, 130, 132,
        135, 138–9, 140–1, 162
    symmetric double-well potential, 190–1
  for s-electron in potential with strong Coulomb
    singularity, 162
  obtained exactly by determination of base function
    such that higher-order contributions vanish,
        140–2
  optimized in first-order approximation by requiring
    that first- and third-order results coincide, 142
  related to scattering of waves by elliptic cone,
        115–17
quasi-stationary state,
  boundary condition for, 30–2, 193
  half-width of, 196
  quantization condition for, 195

Rayleigh–Schrödinger perturbation theory, 197
reflectionless potential, 62–3
removal of boundary condition for quasi-stationary
    state from real axis to anti-Stokes line, 30–2
resonances, 57
  displacement of due to omission of small quantity,
        58
rule of thumb for tracing phase-integral solution
    along path of strict monotonicity, in particular
    Stokes line, 68–9
Rydberg–Klein–Rees procedure, 147

scattering of waves by elliptic cone, 115
splitting of energy levels, *see* distance between
    nearlying energy levels
Stark effect of hydrogen atom treated by
    phase-integral method, great
    accuracy of, 196–7

Stokes and anti-Stokes lines,
  definition of, 28, 63
  emerging from transition point of odd order,
    definition of, 29, 70, 171
  orthogonality of, 28
  simple and double arrows on, 29, 31
Stokes constants associated with,
  complex barrier, 171–5
  simple transition zero, 69–71
  simplest non-monotonic path, 28
Stokes phenomenon, 21
  rigorous method for mastering, 22
supplementary quantity
  ϕ associated with complex single-hump barrier, 11
    formulas for, obtained by comparison equation
        technique, 51, 180
  $\tilde{\phi}$ associated with real single-hump barrier, 11, 47
    formulas for, 51–2, 184–5
symmetry relations for F-matrix elements, 25–6,
    36–7

tracing of wave function,
  along anti-Stokes line, 66
  along path of strict monotonicity, in particular
    Stokes line, 66–9
  rule of thumb for, 68–9
traffic rules for vehicle consisting of compound
    symbol, 30–1
transition point,
  definition of, 66
transition zero,
  definition of, 72
  simple zeros of phase integrand $q(z)$ in
    neighbourhood of, 17
transmission coefficient for real
  Eckart–Epstein barrier, 199
  parabolic barrier, 199–200
  single-hump barrier, 54, 198
    obtained accurately in higher-order
        approximation also for thin barrier, 199
  symmetric double-hump barrier, 203–4
turning point,
  definition of, 35
  in generalized sense, 35
  well-isolated, 35

value of normalized s-electron wave function at the
    origin, 163
value of wave function at turning point, 85, 88
  accuracy of formulas for, 88–91
vehicle consisting of compound symbol, 30
  traffic rules for, 30–1
virial theorem, 168–9

wave function,
  for s-electron in radial potential with strong
    Coulomb singularity, 159–60
  on opposite sides of real barrier, 49, 52–7
  on opposite sides of turning point, 43–5
  represented by phase-integral approximation along
    anti-Stokes line, 66

wave function (*cont.*)
  represented by phase-integral approximation along
    path of strict monotonicity, in particular
    Stokes line, 66–9
wave function associated with real barrier, 49, 52–3,
    183–5
  change of phase and amplitude of on one side
    of the barrier due to change of phase
    on the other side of the barrier,
    55–6
  given as outgoing wave, 53
  given as standing wave, 54–6

  phase of, when its logarithmic derivative is given in
    middle of symmetric barrier, 110–11
  resonance structure of, 57
WKB approximation,
  inadequacy of the name, 8
  relation to earlier known results, 2–7
WKB functions, deficiences of, 13, 23
Wronskian of
  phase-integral functions, 12–14, 20, 23
    importance of its constancy, 13–14
  WKB functions,
    drawbacks of its non-constancy, 13, 23